Build Your Own Hot-Air Balloon

Appendices

by

Eagle Balloons
and
Robert J. Rechs

ULTRA-LIGHT BALLOONS
PREFACE

If money to complete this project safely, is a potential handicap, then PLEASE STOP NOW. If you feel you have the money to do it right, PLEASE DO NOT COMPROMISE SAFETY.

DISCUSSION:

By the definition "Ultra-Light", we are limited to very specific FAA Regulations (FAR-103). Other regulations are particularly applicable (FAR's 61 & 91), and must be adhered to also.

This section will attempt to keep our design limitations as:

1) The balloon, without pilot & fuel, must not exceed 155#;
2) The fuel supply can not exceed 5 Gallons;
3) The balloon must accomodate the pilot only (NO passengers).

PARAMETERS:

From the above limitations, we are generally talking about:

1) Lift - sufficient to carry a 170-200 lb (77-91 kg.) person;
2) on an ambient temperature day of 70-80 F (20-25 C);
3) for 30-40 minutes.

Thus one can see that if the Propane is limited to 5 gallons, the ready-to-fly aircraft without fuel, should not exceed 125 pounds in weight. (155-30#)

ASSUMPTIONS:

An 900 cubic meter balloon can easily be built within these limitations with conventional Nylon fabric. If however, you want a lighter or more compact balloon, piorities must be placed on equipment features, and flight technique must be weighed against safety.

The substitution of one or all of the following must be considered:

1) Use of Dacron, Nomex, Kevlar, or more exotic fabrics; (and using several kinds at various heat locations);
2) No deflation port, vs. a "dump system" (also saves the use of Velcro that won't sustain high temps);
3) No maneuvering vent, to retain maximum heat (requires expert flying technique);
4) One equatorial load-tape, but small 1/2" tape inside seam, or NO vertical load-tapes (to reduce considerable weight)
5) Comfortable pilot accomodations, using a small swing seat or parachute harness;
6) The possibility of incorporaing the fuel tank as part of the pilot accomodations.

FABRICS:

Generally, you can find (or have made) most any fabric you can pay for. But since some of the exotic fabrics available can cost in excess of $100 per yard (in no less than 10,000 yard lots), lets try to stay practical. Availability & price will vary within the trade, but generally:

1) NYLON = 250-300 F = readily available at $2-3.00/yard;
2) DACRON = 300-350 F = generally available at $3-4.00/yard:
3) NOMEX = 450-500 F = available, hard to find $10-20/yard;
4) GENTEX = 550-700 F = hard to find, and $40-50/yard;
5) KEVLAR 750+ °F, very hard to find, $100/yard.

Since you are going to need fabric in the following approximate quantities, consider a multi-fabric envelope:

AX-1 (250cm)	= 233.5 sq. yds	(440 yards x 36" wide)
AX-2 (400cm)	= 319.5 sq. yds	(515 yards x 39" wide)
AX-2.5 (500cm)	= 369.0 sq. yds	(551 yards x 42" wide)
AX-3 (600cm)	= 418.5 sq. yds	(589 yards x 45" wide)
AX-3.3 (700cm)	= 462.0 sq. yds	(629 yards x 48" wide)
AX-3.6 (800cm)	= 505.0 sq. yds	(771 yards x 52" wide)
AX-4 (900cm)	= 548.5 sq. yds	(*).

The maximum stock material width from a wholesaler is 60". Wider material is available from the manufacturer, but only in 10,000 yards of one color. So if the material you purchase is narrower than that shown, you will have to cut your pattern into multiple horizontal gores. It may reduce the scrap-loss, but it is time consuming in sewing, and increases the overall weight.

PURCHASE LIST (partial):

qty/nomenclature	est$	qty/nomenclature	est$
(a) yards fabric		1 Burner, Red-Dragon	
(b) yards load-tape		1 Control handle/pilot	
10 yards 2" webbing		1 pr Seat belts	
4# thread, V69 Dacron		100' Cable, SST x 3/16"	
32 D-Rings		32 Thimbles	
10' Hose, LPG		2 Fittings, LPG brass	
1 Duffle Bag		1 Seat, plastic seat	
32 Snap hooks		10 Misc. bolts	
2 Carabiners, 2"		10' Alum. Angle, 1" x 1"	
sub-total	$2368+	sub-total	+ $348

$2368+ $348 = $$2716.00 MINIMUM

LIFT vs. TEMPERATURE

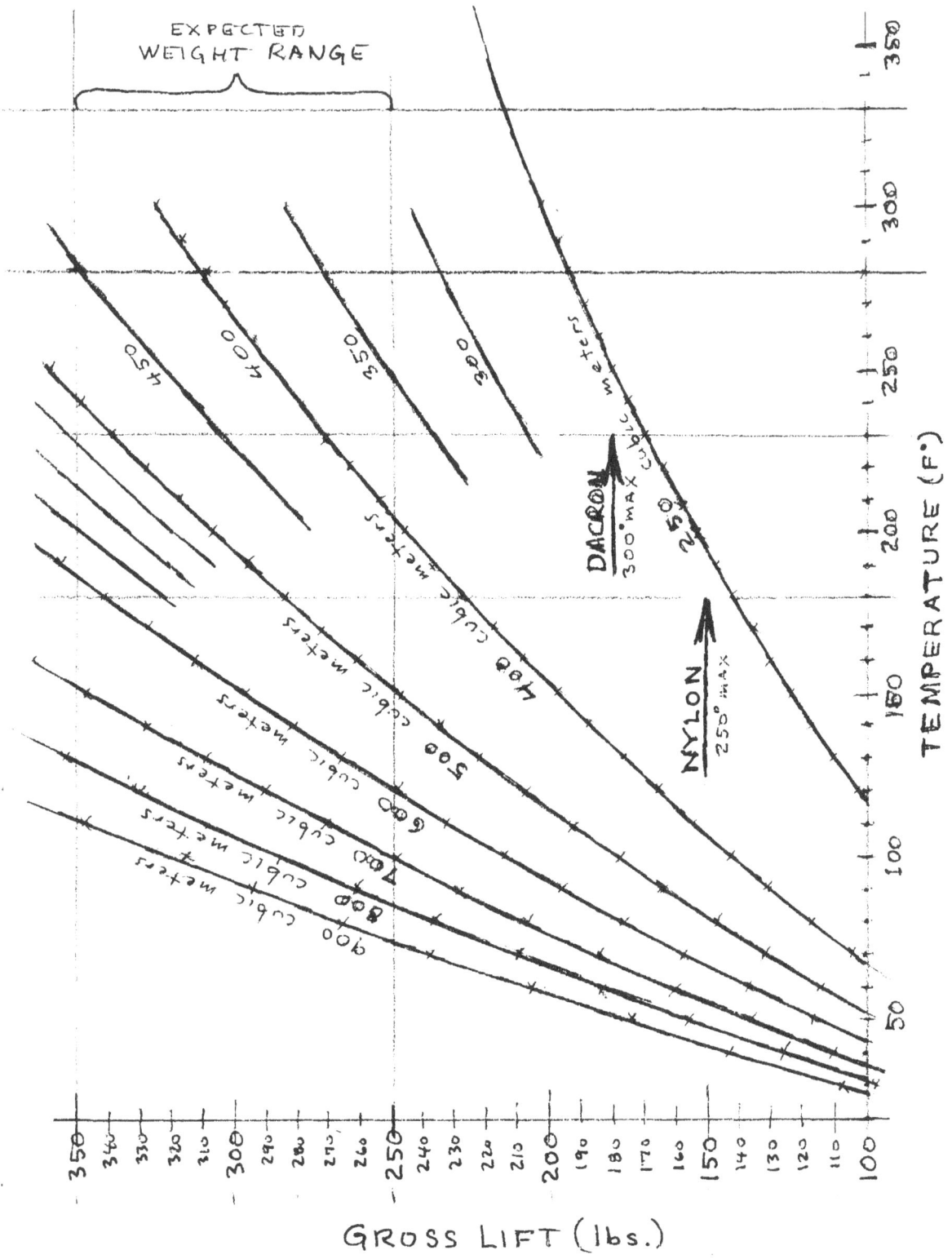

EXPECTED WEIGHT RANGE

DACRON 300° MAX

NYLON 250° MAX

450, 400, 350, 300, 250 cubic meters

400 cubic meters

500 cubic meters

600 cubic meters

700 cubic meters

800 cubic meters

900 cubic meters

TEMPERATURE (F°)

GROSS LIFT (lbs.)

ULTRA-LIGHT BALLOONS

ENVELOPE MATERIAL
Comparisons

trade-name		safe	deforms	melts	notes
ZANTREL		< 150		300	
POLYPROP		<	285-300	320-350	Monofiliment
AVRIL		<		350-400	
LYCRA		<	300	446	
VISCOSE		<	300	350-464	
NYLON-6	*	< 250	300	419-430	poor in ultraviolet
ACETATE		<	350-375	400-445	
TEFLON		<	400	550	hot gas is poison
CLOSPAN		<	420		
ZEFRAN		<	420-490		
CRESLAN		<	430-450		
NUMA		<	437-446	511-518	
DACRON	*	< 300	440-446	482	twice N-66 cost
NYLON-66	*	<	445	480-500	commonly used
ENCRON		<	445	482	
ZEFRAN		<		485	
QIANA		<	447	525	
AVLIN		<	450-468	480-490	
ORLON	*	<	455		bad in sunlight
SPECTRAN		<	455	480-490	
TREVIRA		<	455-465	495	
QUINTESS		<	460	500	
FORTREL		<		478-490	
KODEL-200		<		554-565	
KYNOL		<		572-1080	
ARNEL		<		572	
NOMEX	*	< 500		700	
KEVLAR	*	< 650		840	
GLASS	*	<1500	1560-1778	2050-2160	no fold strength

Notes: * Good for balloons & commercially available.

ULTRA-LIGHT BALLOONS
an excerpt from
AIRLINE MAGAZINE

If balloons are your fancy you can float gently into the sky in a Colt Cloudhopper or Drifter. After inflating the balloon with a propane fueled burner and gasoline powered fan, the pilot straps himself into a harness attached to a small seat. The harness is suspended below the balloon on a swivel so that a slight twist of the body turns the pilot in a different direction. The direction of flight, however, depends on the wind.

Like the ParaPlane, the balloon's controls are simple. Pulling on the burner lever above the head sends a blast of hot air into the balloon, increasing lift. The ascent or descent of the balloon is determined by how frequently the burner is ignited. After landing, the pilot can fully deflate the balloon by pulling on a cord attached to a flap at the top of the envelope.

The only other control is an electronic igniter for the burner. Pushing a button on the shoulder bracket bends a crystal, which generates enough voltage to create a spark. Poof! The pilot light is lit. The burner is specially designed to emit a pencil thin flame to ease inflation and avoid burn damage around the mouth of the balloon. Other features include a buzzer which warns when the fuel is low and a device for measuring the temperature at the top of the envelope.

One of the most remarkable sensations about balloon flight is its serenity. When the burner is not on it is unbelievably quiet. The pilot can hear dogs barking, and people saying Hello. It seems like everyone wants to say Hello to a balloonist.

The Cloudhopper is made by Colt Balloons of Shropshire, England and is marketed in the United States by Early Winters of Seattle, Washington. Cost is about $5,000. A virtually identical balloon called the Drifter is also made by Colt and marketed by Hoverair of San Francisco. It costs about $6,000 which includes a 150 page instruction book, videotape, and several other pieces of equipment. Early Winters and Hoverair say their balloons can be operated as unpowered ultralight aircraft.

In this age of high technology and lots of rules and regulations, there's nothing like going back to the basics for the sheer enjoyment of flying. And the Cloudhopper, the Drifter, and the ParaPlane will provide plenty of that.

ULTRA-LIGHT SIZE
250 cubic meters (8827 cubic feet)
in 32 ONE-PIECE Gores

ENGLISH	dimensions	METRIC
494.838 Inches	Gore length	12.569M
31.764 Inches	Maximum panel width	.806M
32.552 Feet	Inflated Height	9.922M
26.961 Feet	Inflated Diameter	8.281M
439.85 Yards	36"Wide material needed	402.3M
233.5 Sq.Yards	Total surface area:	195.25 M2
63.3 Yards	Load Tape Required:	56.6M
47.5 Feet	Circumference:	43.4M
3 Feet	*Mouth Diameter not smaller than:*	1-Meter
None	Manuvering Vent/Deflation Port	None

```
         INCHES                      TOP                  METERS
    -S-dimension-W-                                   -S-dimension-W-
   494.838"!   0.0  ----------------------*----------------12.569m !  .000
   484.942 !   1.955                                        12.318  !  .049
   475.045 !   3.884                                        12.067  !  .099
   465.148 !   5.829                                        11.815  !  .148
   455.251 !   7.774                                        11.563  !  .197
   445.355 !   9.704                                        11.313  !  .246
   430.509 !  12.574                                        10.935  !  .319
   415.664 !  15.439                                        10.558  !  .392
   400.819 !  18.155                                        10.181  !  .461
   385.974 !  20.867                                         9.804  !  .530
   371.129 !  23.243                                         9.427  !  .590
   356.284 !  25.618                                         9.050  !  .650
   336.490 !  28.166                                         8.547  !  .715
   316.697 !  30.081                                         8.044  !  .764
   269.903 !  31.289                                         7.541  !  .795
   277.110 !  31.764------------+--EQUATOR---+---------      7.039  !  .807
   257.316 !  31.541                                         6.536  !  .801
   437.522 !  30.695                                         6.033  !  .780
   217.729 !  29.319                                         5.531  !  .745
   197.935 !  27.513                                         5.028  !  .699
   183.090 !  24.782                                         4.651  !  .629
   168.245 !  24.193                                         4.274  !  .615
   153.400 !  22.278                                         3.897  !  .565
   138.555 !  20.363            <----W----->-----+----       3.519  !  .517
   123.710 !  18.289                                         3.141  !  .465
   108.864 !  16.216                                         2.765  !  .411
    94.019 !  14.063                                         2.388  !  .357
    79.174 !  11.881                                         2.011  !  .301
    69.277 !  10.411                                         1.759  !  .264
    59.381 !   8.932        *                     S          1.508  !  .227
    49.484 !   7.452 (=75.9")                                1.257  !  .189
    39.587 !   5.968 (=60.8")                                1.006  !  .152
    29.690 !   4.473 (=45.5")                                 .754  !  .114
    19.794 !   2.984 (=30.4")                                 .502  !  .076
     9.897 !   1.489 (mouth dia.)                             .251  !  .038
     0.000--- 0.000--------------------*-----------+-----      .000 ---  .000
    -S-dimension-W-                MOUTH END                -S-dimension-W-
```

NOTE: The above are ONE-piece gores. ADD 1" to "W" for sew-seams.
    Do NOT attempt to change the number of gores by varing the "W"
dimension. For anything other than a 32 gore balloon, recompute
all dimensions using VOLUME-I APPENDIX-C.

ULTRA-LIGHT SIZE
400 cubic meters (14124 cubic feet)
in 32 ONE-PIECE Gores

ENGLISH	dimensions	METRIC
578.767 Inches	Gore length	14.701M
37.151 Inches	Maximum panel width	.944M
37.1 Feet	Inflated Height	11.3M
31.5 Feet	Inflated Diameter	9.6M
514.5 Yards	39"Wide material needed	470.4M
319.4 Sq.Yards	Total surface area:	267 M2
61.3 Yards	Load Tape Required:	61.3M
99 Feet	Circumference:	31.5M
3 Feet	*Mouth Diameter not smaller than:*	1.5Meters
None	Manuvering Vent/Deflation Port	None

```
       INCHES                   TOP              METERS
    -S-dimension-W-                           -S-dimension-W-
    578.767"-- 0.0 --------------------------14.701m--- .000
    567.191 !  2.286                          14.407  !  .058
    555.616 !  4.543                          14.113  !  .115
    544.041 !  6.818                          13.819  !  .173
    532.465 !  9.092                          13.525  !  .230
    520.890 ! 11.350                          13.231  !  .288
    503.526 ! 14.706                          12.790  !  .374
    486.164 ! 18.058                          12.349  !  .459
    468.801 ! 21.235                          11.908  !  .539
    451.438 ! 24.407                          11.467  !  .620
    434.075 ! 27.185                          11.026  !  .691
    416.712 ! 29.963                          10.584  !  .761
    393.561 ! 32.943                           9.997  !  .837
    370.411 ! 35.183                           9.409  !  .894
    347.260 ! 36.595                           8.821  !  .929
    324.109 ! 37.151----------EQUATOR--------- 8.233  !  .944
    300.959 ! 36.890                           7.645  !  .937
    277.808 ! 35.901                           7.057  !  .912
    254.657 ! 34.292                           6.469  !  .871
    231.507 ! 32.179                           5.881  !  .817
    214.144 ! 28.985                           5.439  !  .736
    196.781 ! 28.296                           4.998  !  .719
    179.418 ! 26.056                           4.557  !  .612
    162.055 ! 23.816    <----W----><----+----  4.116  !  .605
    144.692 ! 21.391                           3.675  !  .543
    127.329 ! 18.966                           3.234  !  .481
    109.966 ! 16.449                           2.793  !  .418
     92.603 ! 13.896                           2.352  !  .353
     81.027 ! 12.177                           2.058  !  .309
     69.452 ! 10.447        *            S     1.764  !  .265
     57.877 !  8.716  (=88.8")                 1.470  !  .221
     46.301 !  6.980  (=71.1")                 1.176  !  .177
     34.726 !  5.232  (=53.3")                  .882  !  .133
     23.151 !  3.490  (=35.5")                  .588  !  .089
     11.575 !  1.742  (mouth dia.)              .294  !  .044
      0.000--- 0.000--------------------------- .000 --- .000
    -S-dimension-W-            MOUTH END       -S-dimension-W-
```

NOTE: The above are ONE-piece gores. ADD 1" to "W" for sew-seams.
    Do NOT attempt to change the number of gores by varing the "W"
dimension. For anything other than a 32 gore balloon, recompute
all dimensions using VOLUME-I APPENDIX-D.

```
                    ULTRA-LIGHT SIZE
          500 cubic meters (17,655 cubic feet)
                in 32   ONE-PIECE Gores
ENGLISH                    dimensions                     METRIC
623.457 Inches         Gore length:                      15.837M
40.020 Inches       Maximum panel width:                  1.017M
39.5 Feet             Inflated Height:                     12.1M
34.0 Feet            Inflated Diameter:                    10.4M
554.2 Yards       42"Wide material needed:                506.8M
370.6 Sq.Yards       Total surface area:                  310.M2
70.68 Yards         Load Tape Required:                    63.5M
106 Feet              Circumference:                       32.4M
4 Feet        *Mouth Diameter not smaller than:*       1.2Meters
None          Deflation Port/Manuvering Vent:               None
```

```
         INCHES                                      METERS
    -S-dimension-W-              TOP              -S-dimension-W-
   623.457"!   0.0  ------------- * -----------15.836m !  .000
   610.988 !   2.463                             15.519 !  .063
   598.519 !   4.894                             15.203 !  .124
   586.050 !   7.344                             14.886 !  .187
   573.581 !   9.795                             14.570 !  .249
   561.111 !  12.226                             14.253 !  .311
   542.408 !  15.842                             13.778 !  .402
   523.704 !  19.451                             13.303 !  .494
   505.001 !  22.875                             12.828 !  .581
   486.297 !  26.291                             12.352 !  .668
   467.593 !  29.284                             11.877 !  .744
   448.889 !  32.276                             11.402 !  .820
   423.951 !  35.487                             10.769 !  .901
   399.013 !  37.900                             10.135 !  .963
   374.074 !  39.421                              9.502 ! 1.001
   349.136 !  40.020--------------EQUATOR---+---- 8.868 ! 1.017
   324.198 !  39.739                              8.235 ! 1.009
   299.260 !  38.673                              7.602 !  .982
   274.321 !  36.940                              6.968 !  .938
   249.383 !  34.664                              6.335 !  .881
   230.679 !  31.223                              5.859 !  .793
   211.975 !  30.481                              5.384 !  .774
   193.272 !  28.068                              4.909 !  .713
   174.568 !  25.655          <----W---->---+----  4.434 !  .652
   155.864 !  23.043                              3.959 !  .585
   137.161 !  20.431                              3.484 !  .519
   118.457 !  17.719                              3.009 !  .450
    99.753 !  14.969                              2.534 !  .380
    87.284 !  13.118                              2.217 !  .333
    74.815 !  11.253                    S         1.900 !  .286
    62.346 !   9.389       *                       1.584 !  .238
    49.877 !   7.519=(76.6")                       1.267 !  .191
    37.407 !   5.636=(57.4")                        .950 !  .143
    24.938 !   3.759=(38.3")                        .633 !  .095
    12.469 !   1.877=(mouth dia.)                   .315 !  .047
     0.000--- 0.000------------------------+----    .000 ---  .000
    -S-dimension-W-            MOUTH END         -S-dimension-W-
```

NOTE: The above are ONE-piece gores. ADD 1" to "W" for sew-seams.
     Do NOT attempt to change the number of gores by varing the "W"
dimension. For anything other than a 32-gore balloon, recompute
all dimensions using VOLUME-I APPENDIX-D.

ULTRA-LIGHT SIZE
600 cubic meters (21,186 cubic feet)
in 32 ONE-PIECE Gores

ENGLISH	dimensions	METRIC
662.522 Inches	Gore length:	16.829M
42.527 Inches	Maximum panel width:	1.080M
41.5 Feet	Inflated Height:	12.7M
36.1 Feet	Inflated Diameter:	11.0M
588.9 Yards	45"Wide material needed:	538.5M
418.5 Sq.Yards	Total surface area:	350.M2
73.0 Yards	Load Tape Required:	None
113.4 Feet	Circumference:	34.6M
5 Feet	*Mouth Diameter not smaller than:*	1.5Meters
None	Manuvering Vent/Deflation Port:	None

```
        INCHES                        TOP               METERS
   -S-dimension-W-                                   -S-dimension-W-
   662.522"!   0.0  --------------------------------16.828m--- .000
   649.272 !   2.617                                 16.492  !  .066
   636.021 !   5.201                                 16.156  !  .132
   622.771 !   7.805                                 15.819  !  .198
   609.521 !  10.408                                 15.483  !  .264
   596.270 !  12.992                                 15.146  !  .330
   576.394 !  16.835                                 14.641  !  .428
   556.519 !  20.671                                 14.136  !  .525
   536.643 !  24.308                                 13.631  !  .617
   516.767 !  27.939                                 13.126  !  .710
   496.892 !  31.119                                 12.622  !  .790
   477.016 !  34.299                                 12.117  !  .871
   450.515 !  37.711                                 11.444  !  .958
   424.014 !  40.275                                 10.770  ! 1.023
   397.513 !  41.891                                 10.097  ! 1.064
   371.013 !  42.527----------+---EQUATOR----+---------  9.424  ! 1.080
   344.512 !  42.229                                  8.751  ! 1.073
   318.011 !  41.096                                  8.078  ! 1.044
   291.510 !  39.254                                  7.405  !  .997
   265.009 !  36.836                                  6.732  !  .936
   245.133 !  34.617                                  6.227  !  .843
   225.257 !  32.391                                  5.722  !  .823
   205.382 !  29.827                                  5.217  !  .758
   185.506 !  27.263        <----W---->------+---      4.712  !  .693
   165.631 !  24.487                                  4.207  !  .622
   145.755 !  21.701                                  3.702  !  .551
   125.879 !  18.829                                  3.197  !  .478
   106.004 !  15.907                                  2.693  !  .404
    92.753 !  13.939                                  2.356  !  .354
    79.503 !  11.959                        S         2.019  !  .304
    66.252 !   9.978       *                          1.683  !  .253
    53.002 !   7.990=(81.4")                          1.346  !  .203
    39.751 !   5.989=(61.0")                          1.010  !  .152
    26.501 !   3.995=(40.7")                           .673  !  .101
    13.250 !   1.994 (mouth dia.)                      .337  !  .051
     0.000--- 0.000-------------------+---------+---   .000 --- .000
   -S-dimension-W-            MOUTH END             -S-dimension-W-
```

NOTE: The above are ONE-piece gores. ADD 1" to "W" for sew-seams.
    Do NOT attempt to change the number of gores by varing the "W"
dimension. For anything other than a 32-gore balloon, recompute
all dimensions using VOLUME-I APPENDIX-D.

ULTRA-LIGHT SIZE
700 Cubic Meters (24717 cubic feet)
in 32 ONE-PIECE Gores

ENGLISH	dimensions	METRIC
697.455 Inches	Gore length:	17.716M
44.770 Inches	Maximum panel width:	1.137 M
43.4 Feet	Inflated Height:	13.2 M
38.0 Feet	Inflated Diameter:	11.6 M
620.0 Yards	48"Wide material needed:	566.9 M
463.8 Sq.Yards	Total surface area:	387.8M2
75.1 Yards	Load Tape Required:	67.6 M
119.4 Feet	Circumference:	36.4 M
72 Inches	*Mouth Diameter not smaller than:*	2 Meters
None	Manuvering Vent/Deflation Port:	None

```
       INCHES                                              METERS
  -S-dimension-W-                    TOP              -S-dimension-W-
  697.455"!   0.0 ---------------------------------17.716M !   .0 M
  683.506 !   2.755                                 17.362  !  .070
  669.557 !   5.475                                 17.007  !  .139
  655.608 !   8.216                                 16.653  !  .209
  641.659 !  10.957                                 16.299  !  .278
  627.709 !  13.677                                 15.945  !  .347
  606.786 !  17.722                                 15.413  !  .450
  585.862 !  21.761                                 14.882  !  .553
  564.938 !  25.590                                 14.350  !  .650
  544.015 !  29.412                                 13.819  !  .747
  523.091 !  32.759                                 13.287  !  .832
  502.168 !  36.107                                 12.756  !  .917
  474.269 !  39.699                                 12.047  ! 1.008
  446.371 !  42.398                                 11.338  ! 1.077
  418.473 !  44.100                                 10.630  ! 1.120
  390.575 !  44.770--------EQUATOR---------          9.921  ! 1.137
  362.677 !  44.456                                  9.212  ! 1.129
  334.778 !  43.263                                  8.504  ! 1.099
  306.880 !  41.324                                  7.795  ! 1.050
  278.982 !  38.778                                  7.086  !  .985
  258.058 !  36.442                                  6.555  !  .887
  237.135 !  34.099                                  6.023  !  .866
  216.211 !  31.399                                  5.492  !  .798
  195.287 !  28.700        <----W---------           4.961  !  .729
  174.364 !  25.778                                  4.429  !  .655
  153.440 !  22.856                                  3.898  !  .581
  132.516 !  19.821                                  3.366  !  .503
  111.593 !  16.745                                  2.835  !  .425
   97.644 !  14.674                                  2.480  !  .373
   83.695 !  12.589            S                     2.126  !  .320
   69.745 !  10.504      *                           1.772  !  .267
   55.796 !   8.411=(85.7")                          1.417  !  .214
   41.847 !   6.305=(64.2")                          1.063  !  .160
   27.898 !   4.206=(42.8")                           .709  !  .107
   13.949 !   2.099 (mouth dia.)                      .354  !  .053
    0.000---  0.000------------------------------     .000 ---  .000
  -S-dimension-W-                 MOUTH END        -S-dimension-W-
```

NOTE: The above are ONE-piece gores. ADD 1" to "W" for sew-seams.
    Do NOT attempt to change the number of gores by varing the "W"
dimension. For anything other than a 32-gore balloon, recompute
all dimensions using VOLUME-I APPENDIX-D.

```
                    ULTRA-LIGHT SIZE
              800 Cubic Meters (28248 cubic feet)
                  in 32   ONE-PIECE Gores
ENGLISH                    dimensions                    METRIC
729.200 Inches          Gore length:                   18.522M
46.807 Inches        Maximum panel width:               1.189 M
45.2 Feet              Inflated Height:                  13.8 M
39.8 Feet              Inflated Diameter                 12.1 M
648.2 Yards        48" Wide material needed:            592.7 M
507.0 Sq.Yards        Total surface area:               423.9M2
77.0 Yards           Load Tape Required:                 69.3 M
124.8 Feet             Circumferenᴵᴵe:                    37.7 M
72 Inches        *Mouth Diameter not smaller than:*    2 Meters
None             Manuvering Vent/Deflation Port:          None
```

```
        INCHES                                     METERS
  -S-dimension-W-               TOP            -S-dimension-W-
  729.200"!   0.0  ------------------------------18.522M !  .0 M
  714.616 !   2.880                             18.152  !  .073
  700.032 !   5.724                             17.782  !  .145
  685.448 !   8.590                             17.411  !  .218
  670.864 !  11.455                             17.041  !  .291
  656.280 !  14.299                             16.670  !  .363
  634.404 !  18.528                             16.115  !  .471
  612.528 !  22.751                             15.559  !  .578
  590.652 !  26.754                             15.003  !  .680
  568.776 !  30.750                             14.448  !  .781
  546.900 !  34.251                             13.892  !  .870
  525.024 !  37.751                             13.336  !  .959
  495.856 !  41.506                             12.595  ! 1.054
  466.688 !  44.328                             11.854  ! 1.126
  437.520 !  46.107                             11.113  ! 1.171
  408.352 !  46.807-----------+---EQUATOR---+------- 10.373  ! 1.189
  379.184 !  46.479                              9.632  ! 1.181
  350.016 !  45.232                              9.891  ! 1.149
  320.848 !  43.205                              8.150  ! 1.097
  291.680 !  40.544                              7.409  ! 1.030
  269.804 !  36.518                              7.853  !  .928
  247.928 !  35.651                              6.298  !  .906
  226.052 !  32.829                              5.742  !  .834
  204.176 !  30.007          <----W---->----↑---  5.186  !  .762
  182.300 !  26.951                              4.631  !  .685
  160.424 !  23.896                              4.075  !  .607
  138.548 !  20.724                              3.519  !  .526
  116.672 !  17.508                              3.964  !  .445
  102.088 !  15.342                              2.593  !  .390
   87.504 !  13.162                      S       2.222  !  .334
   72.920 !  10.982        *                     1.852  !  .279
   58.336 !   8.794=(89.6")                       1.482  !  .223
   43.752 !   6.592=(67.1")                       1.111  !  .167
   29.168 !   4.397=(44.8")                        .741  !  .112
   14.584 !   2.195 (mouth dia.)                   .370  !  .056
    0.000---  0.000------------------+--------+------ .000 ---  .000
  -S-dimension-W-               MOUTH END         -S-dimension-W-
```

NOTE: The above are ONE-piece gores. ADD 1" to "W" for sew-seams.
    Do NOT attempt to change the number of gores by varing the "W"
dimension. For anything other than a 32-gore balloon, recompute
all dimensions using VOLUME-I APPENDIX-D.

ULTRA-LIGHT SIZE
900 Cubic Meters (31779 cubic feet)
in 32   ONE-PIECE Gores

ENGLISH <<<<<<<<<<<<<<<<<<<	dimensions >>>>>>>>>>>>>>>>	METRIC
758.399 Inches	Gore length:	19.264M
48.682 Inches	Maximum panel width:	1.237M
46.7 Feet	Inflated Height:	14.3 M
41.4 Feet	Inflated Diameter	12.6 M
674.2 Yards	52" Wide material needed:	616.5 M
548.4 Sq.Yards	Total surface area:	458.6M2
78.7 Yards	Load Tape Required:	70.8 M
43.3 Yards	Circumference:	39.6 M
72 Inches	*Mouth Diameter not smaller than:*	2 Meters
None	Manuvering Vent/Deflation Port:	None

```
        INCHES                                          METERS
    S-dimension-W-                  TOP            -S-dimension-W-
    758.399 !   0.0"-------------------------------19.264m   .0 M
    743.231 !   2.996                             18.879 !   .076
    728.063 !   5.953                             18.493 !   .151
    712.895 !   8.934                             18.108 !   .227
    697.727 !  11.914                             17.723 !   .303
    682.559 !  14.872                             17.338 !   .378
    659.807 !  19.272                             16.760 !   .490
    637.055 !  23.662                             16.182 !   .601
    614.303 !  27.826                             15.603 !   .701
    591.551 !  31.982                             15.026 !   .812
    568.799 !  35.622                             14.446 !   .905
    546.047 !  39.262                             13.870 !   .997
    515.712 !  43.168                             13.100 !  1.097
    485.376 !  46.109                             12.029 !  1.171
    455.040 !  47.954                             11.559 !  1.218
    424.704 !  48.682-----------+---EQUATOR---+------- 10.788 !  1.237
    394.368 !  48.340                             10.017 !  1.228
    364.031 !  47.043                              9.247 !  1.195
    333.695 !  44.935                              8.476 !  1.141
    303.360 !  42.167                              7.701 !  1.071
    280.608 !  39.626                              7.128 !  1.007
    257.856 !  37.078                              6.550 !   .942
    235.104 !  34.143                              5.972 !   .867
    212.351 !  31.208         <-----W----+----- ---  5.394 !   .793
    189.600 !  28.030                              4.816 !   .712
    166.848 !  24.853                              4.238 !   .631
    144.096 !  21.554                              3.660 !   .547
    121.343 !  18.209                              3.082 !   .463
    106.176 !  15.957                              2.697 !   .405
     91.008 !  13.689                       S      2.312 !   .348
     75.840 !  11.421         *                    1.926 !   .290
     60.672 !   9.146=(93.2")                       1.541 !   .232
     45.504 !   6.856=(69.8")                       1.156 !   .174
     30.336 !   4.573=(46.6")                        .771 !   .116
     15.168 !   2.283 (mouth dia.)                   .385 !   .058
      0.000--- 0.000-----------------------------  .000 !   .000
    -S-dimension-W-             MOUTH END          -S-dimension-W-
```

NOTE: The above are ONE-piece gores. ADD 1" to "W" for sew-seams.
    DO NOT attempt to change the number of gores by varing the "W"
dimension. For anything other than a 32-gore balloon, recompute
all dimensions using VOLUME-I APPENDIX-D.

PILOT STATION

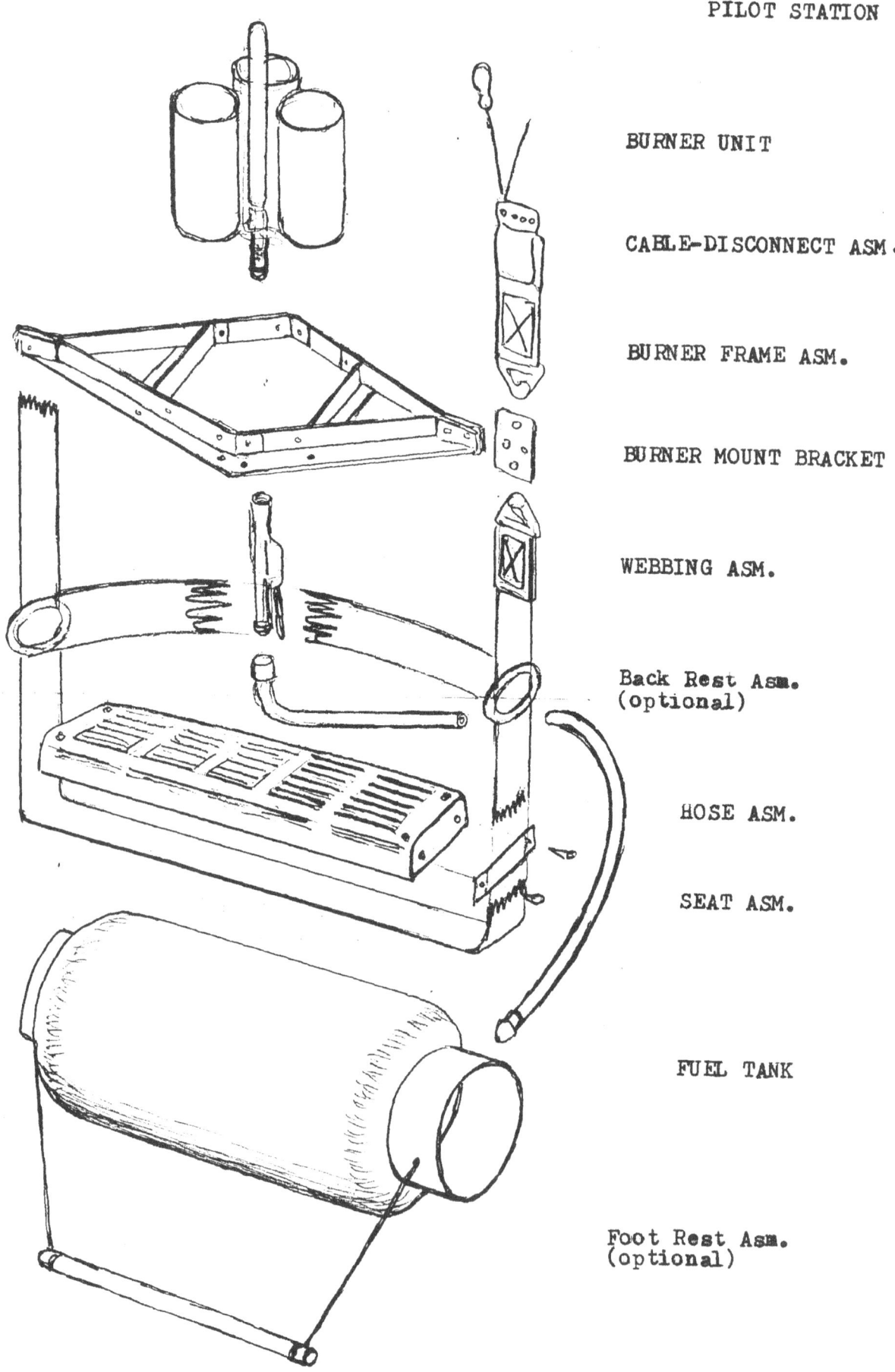

BURNER UNIT

CABLE-DISCONNECT ASM.

BURNER FRAME ASM.

BURNER MOUNT BRACKET

WEBBING ASM.

Back Rest Asm.
(optional)

HOSE ASM.

SEAT ASM.

FUEL TANK

Foot Rest Asm.
(optional)

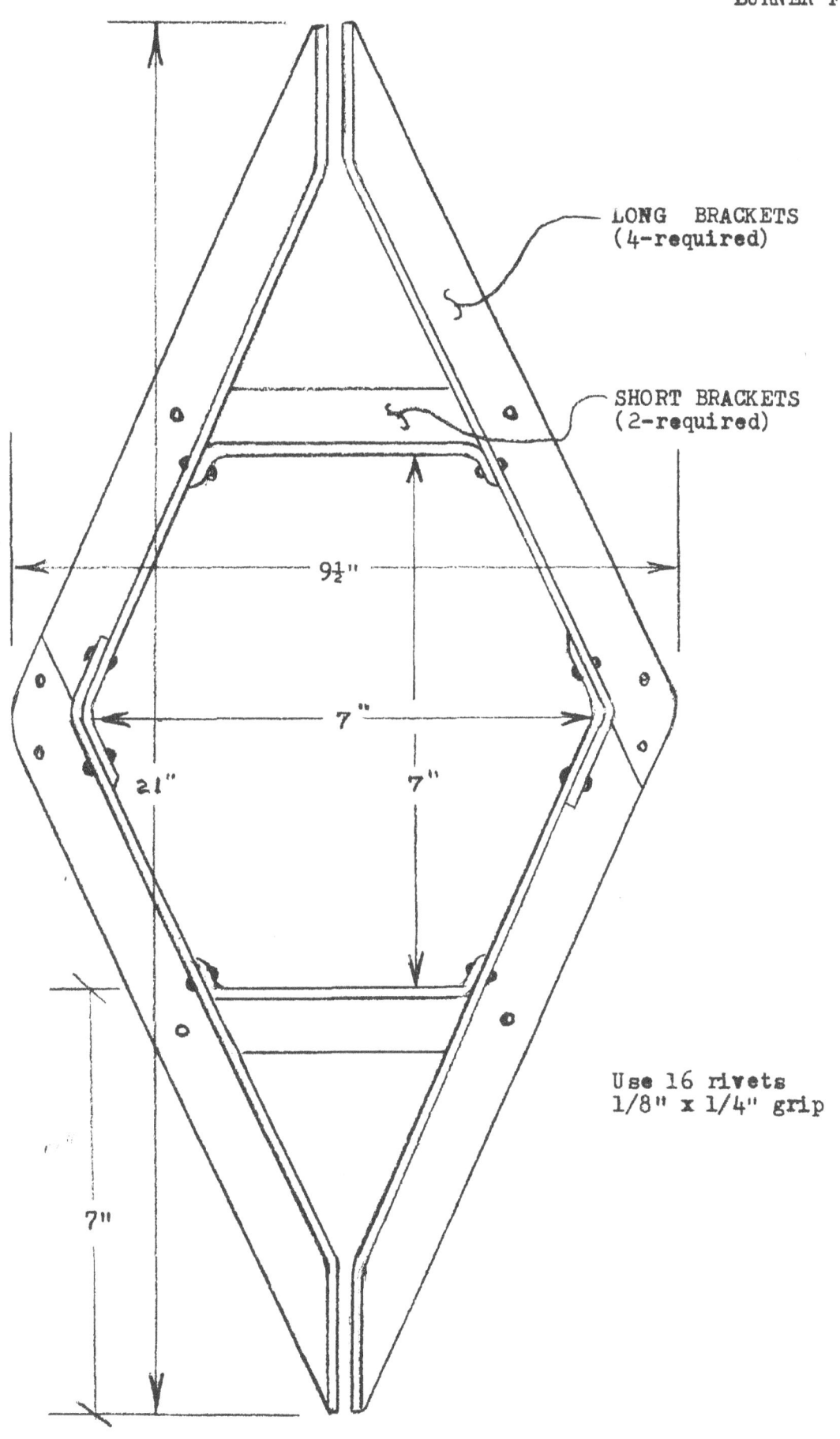

LONG BRACKETS
(4-required)

SHORT BRACKETS
(2-required)

$9\frac{1}{2}$"

7 "

21"

7"

7"

Use 16 rivets
1/8" x 1/4" grip

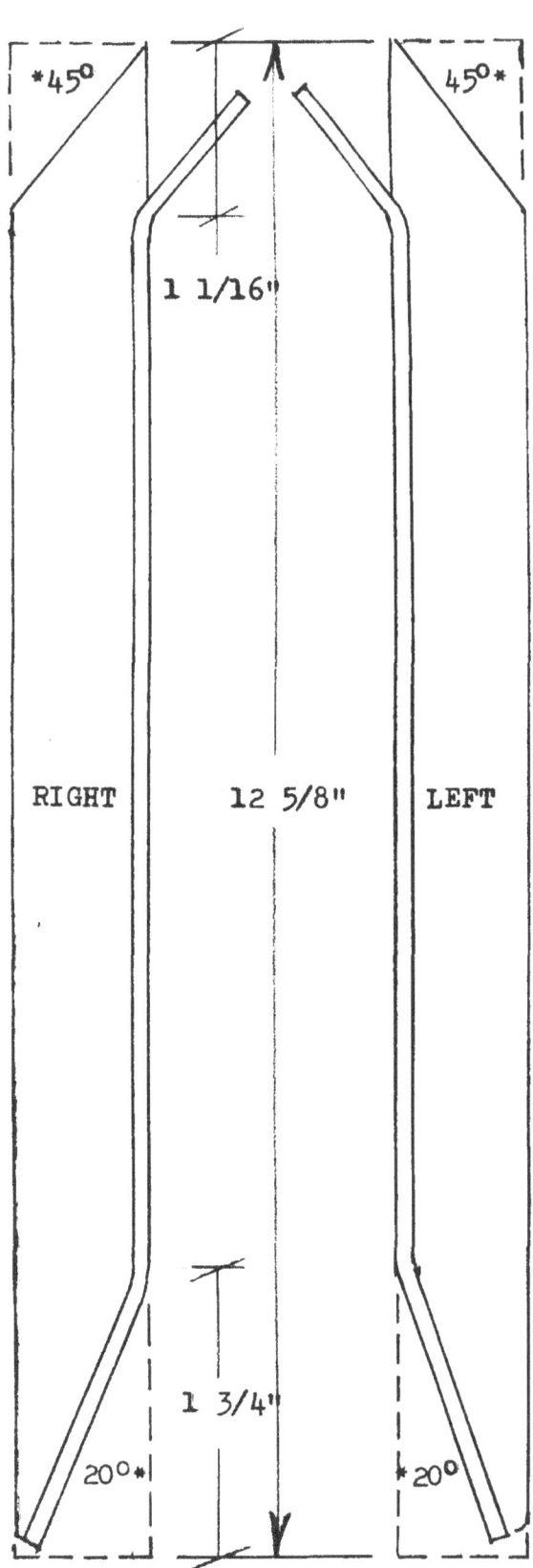

45°   45°

1 1/16"

RIGHT   12 5/8"   LEFT

1 3/4"

20°*   *20°

(TWO of each req.)
MATERIAL: Alum. Angle
SIZE: 1"x1"x1/8" thk.
LENGTH: 12 3/4"
NOTE RIGHT & LEFT
* Trim to fit later

(TWO required)
MATERIAL: Aluminum Angle
SIZE: 1" x 1" x 1/8" thk.
LENGTH: 6 3/4"

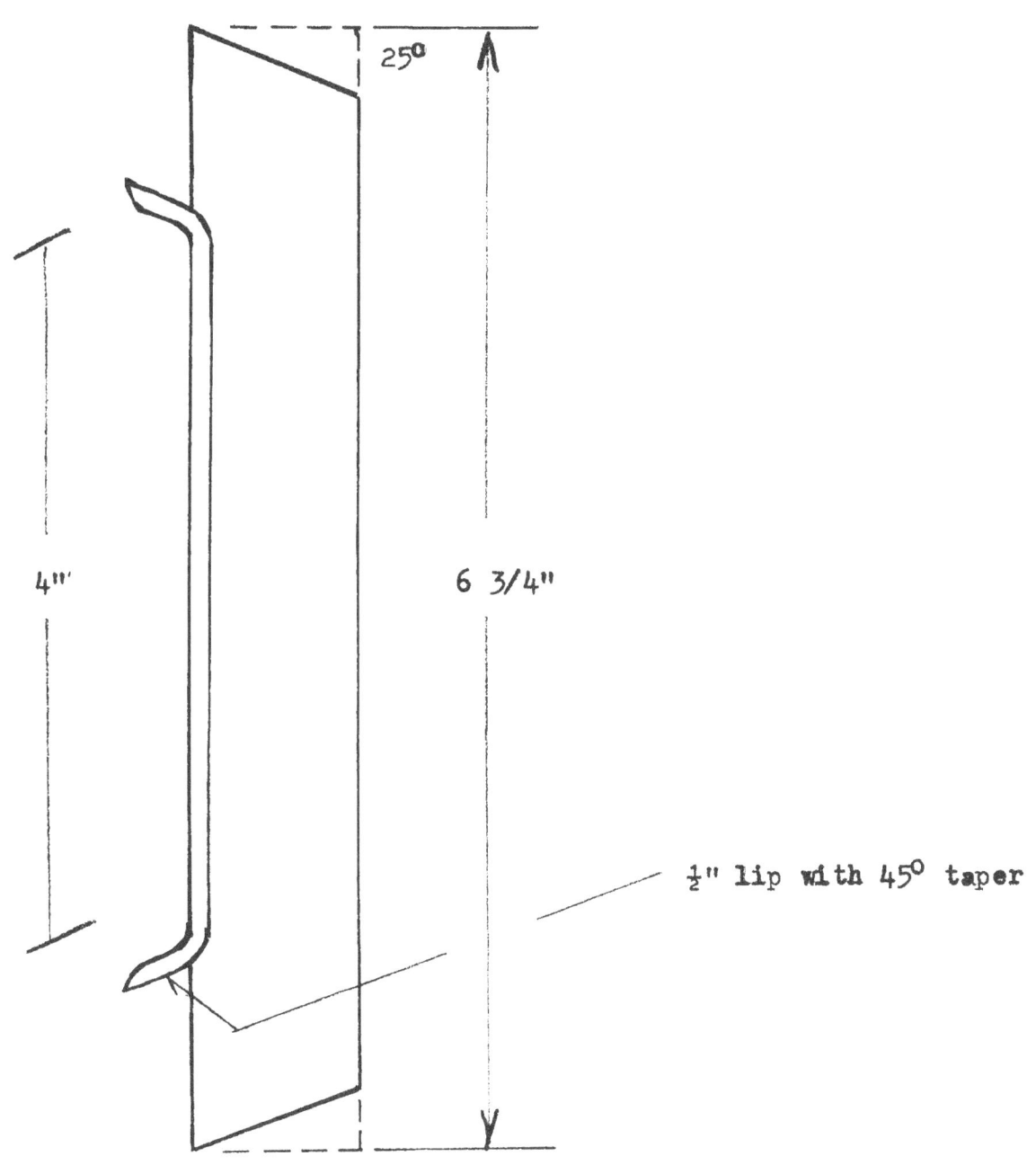

25°

4"

6 3/4"

½" lip with 45° taper

(TWO required)
MATERIAL: Steel plate
SIZE: 2" x 2½ x 3/16 thk.
  (¼" thk. substitute OK)

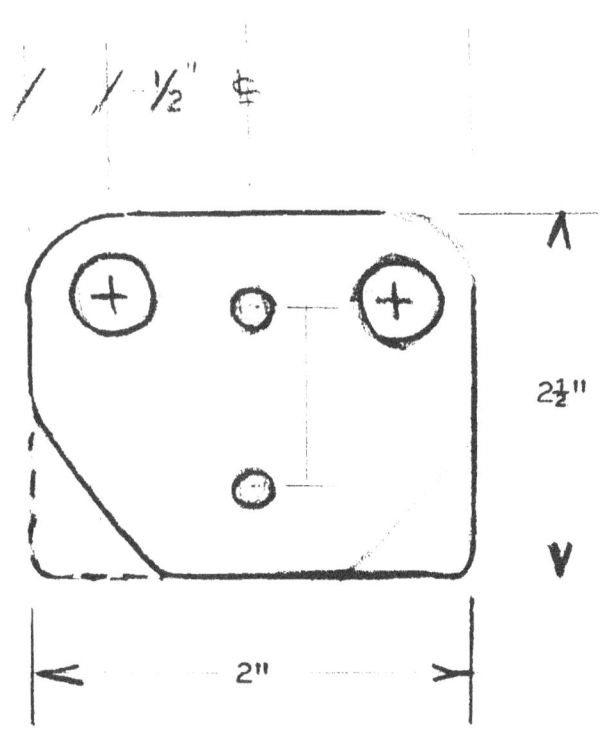

2½"

2"

½" ₵

¼" drill 2-places
  chamfer 45° both sides

3/16 drill 2-places

45° corners optional

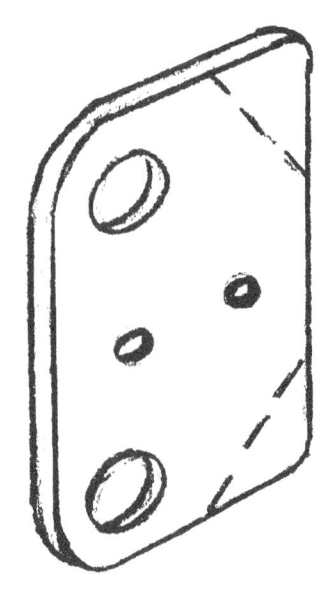

OBLIQUE VIEW

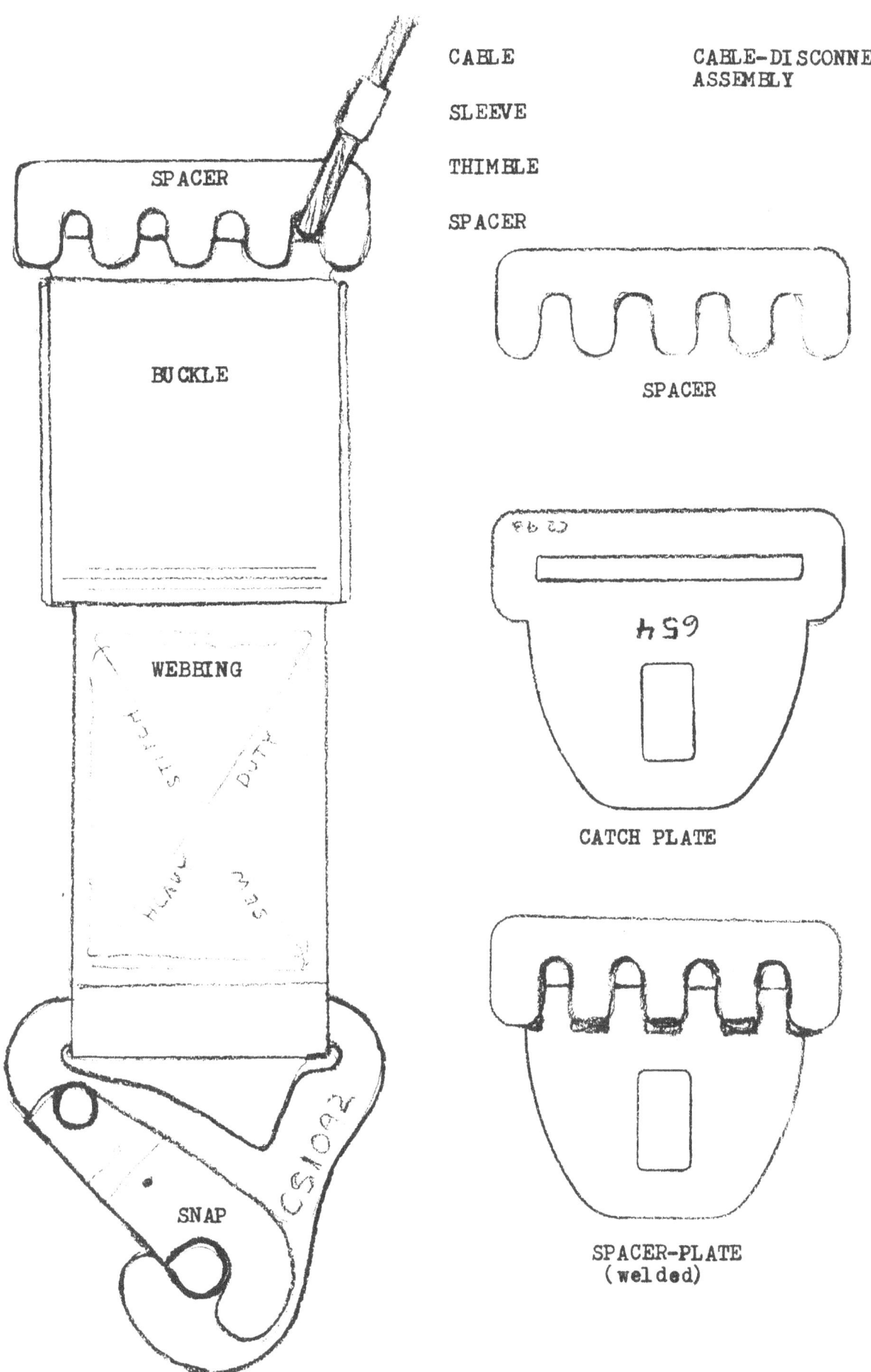

SPACER

BUCKLE

WEBBING

SNAP

CABLE

SLEEVE

THIMBLE

SPACER

CABLE-DISCONNECT
ASSEMBLY

SPACER

CATCH PLATE

SPACER-PLATE
(welded)

SPHERICAL
GAS BALLOONS

NOTES

collected by

ROBERT J. RECHS

Although gas ballooning has shown a small growth over the last decade, it has been a steady growth. Since I originally started collecting this data in the late 1960's, the development of construction has taken some giant leaps forward. While the Europeans cling to the traditional "spherical" envelope made with neoprene coated cotton, the American aeronauts have progressed to "natural" shapes made of plastic films.

These notes are made available in the hopes of preserving the technology and mystique of the "traditional" for future generations of aeronauts. Be aware that many authoritative sources on construction and flying technique are available in selected public libraries.

Those considered the best, all published before 1940, are quite rare, but found in the main city libraries of: Los Angeles, New York City, Chicago, Akron, Ohio.

The two places that undoubtedly have them all are: The Air & Space Museum Library, Washington, DC and The Air Force (Gimble) Library, Colorado Springs, Colorado.

The best books published are:

Airships Past And Present,
Pocket Book of Aeronautics[1],
Free and Captive Balloons[2], by Upson & Chandler,
Balloon And Airship Gases[3],
Through The Air, by J. Wise.

[1] https://archive.org/details/pocketbookofaero00moediala
[2] http://archivale.com/catalog/product_info.php/products_id/1574
[3] http://archivale.com/catalog/product_info.php/products_id/2069

Other technical information on the subject can be found in the files of:

The Lighter Than Air Society
1800 Triplet Blvd
Akron, Ohio 44306
USA

Association of Balloon & Airship Constructors
P.O. Box 3841
La Puente, California
USA
http://abac.archivale.com

Best regards in your research, [former] ISBN # 0-937566-27-9

Robert J. Rechs, author/aeronaut © 1984.

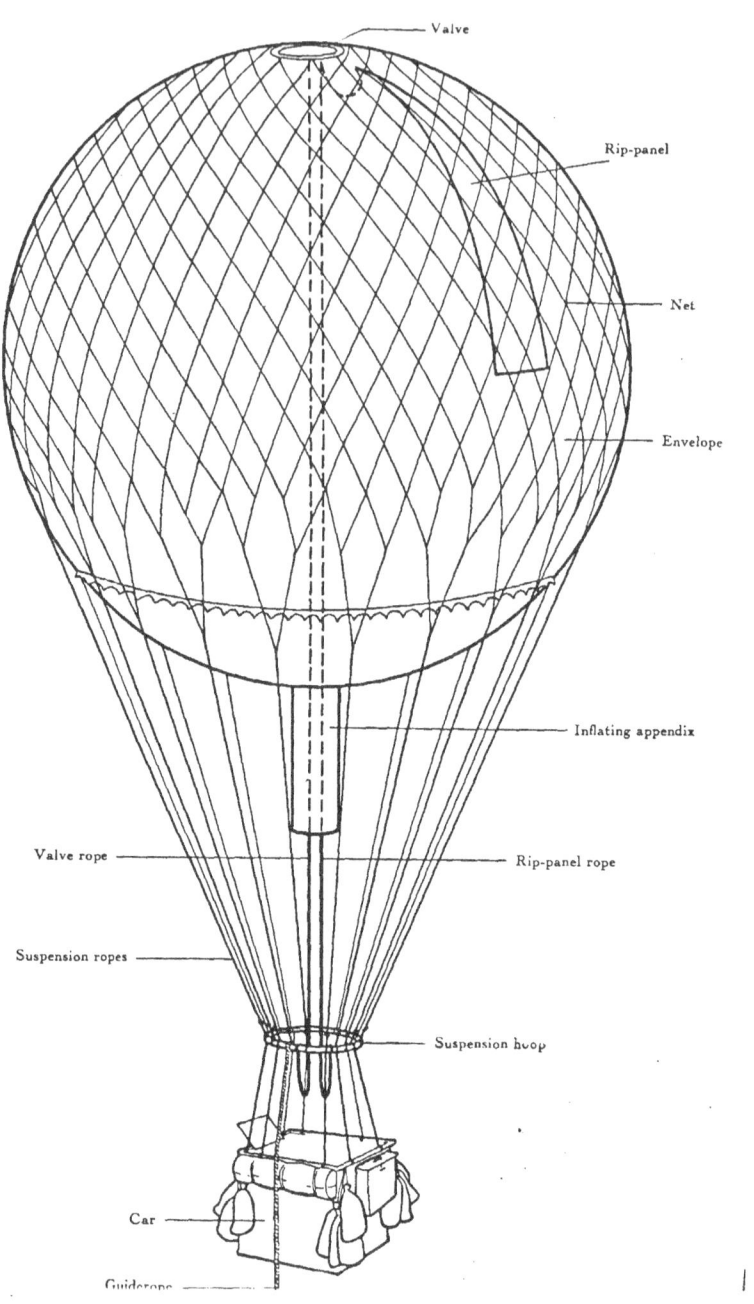

Valve

Rip-panel

Net

Envelope

Inflating appendix

Valve rope

Rip-panel rope

Suspension ropes

Suspension hoop

Car

Guideropes

22

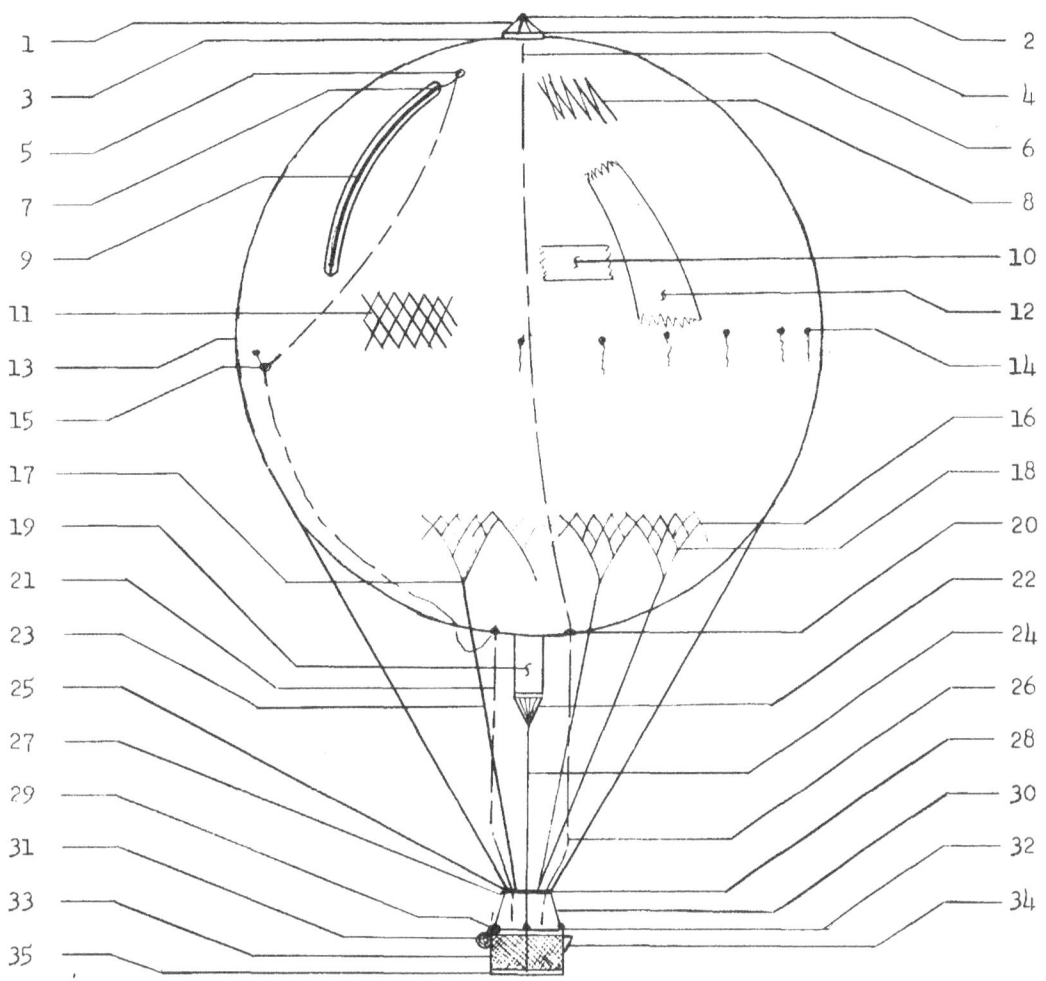

FREE BALLOON - U.S. NAVY -1951

1	Elastic Closers -(6)	12	Gore Position -(12)	24	Appendix Cord	
2	Wishbone	13	Envelope or Bag	25	Load Ring Toggles -(24)	
3	Straps for Net Ring -(8)	14	Parachute Stops -(24)	26	Valve Rope -	
4	Valve, Butterfly Type	15	Rip Cord Gland		white or yellow	
5	Rip Cord Stop -(3)	16	Heavy Net Diamonds	27	Drag Rope-Coiled	
6	Valve, Bridle (6 strand)		(2 Rows of 96)	28	Load Ring	
7	Boots - (2)	17	2nd Crows Feet-(24)	29	Gunwale	
8	Net Diamonds	18	1st Crows Feet-(48)	30	Basket Ropes -(12)	
	(alt. red & blue pcs.)	19	Appendix	31	Drag Rope Toggle & Bridle	
9	Rip Panel	20	Valve Rope Gland	32	Stringers -(6)	
10	Panel Position -(20)	21	Rip Cord- Red	33	Basket	
11	Net equator - solid	22	Appendix Bridle	34	Sand Trough	
	blue (1 row of 96)	23	Foot Ropes -(24)	35	Runners - (4)	

Reprint Courtesy of DON PICCARD BALLOONS

Envelope Design

In determining patterns, the usual assumption is that the fabric[5] can be curved in only [one] direction at a time. Both of the most common patterning systems depend on this same general assumption.

Tho cylindrical gore system consists of a series of gores curved from top to bottom of the balloon, all the same and of a horizontal width commonly governed by the width of the fabric. The width may really be anything however, provided tho number of gores be not less than ten to twelve. There may be any number of transverse seams running in any direction whatever, as long as the sides of tho finished gore maintain tho proper curvature when laid out flat. Tho panels may be staggered, brick fashion, if desired, but in this case they have circular horizontal edges.

In the conical ring system the straight elements of the panels run vertically instead of horizontally. Thus the balloon is built up of a series of horizontal conical frustums, each of which must be calculated separately. In this case the shape of tho finished balloon depends on tho curvature of the horizontal edges, so that theoretically the vertical seams may occur anywhere. Practically, however, there is too much waste of fabric (due to the deep curvature of the horizontal edges) if the number of gores is reduced much below 12 to 16, depending on the size of tho balloon. Therefore, unless the panels are to be symmetrically staggered there is no real excuse for the conical ring system with its complicated patterning and its large waste of fabric.

It should be noted that the above mentioned pattern systems have nothing to do necessarily with the overall dimensions or general arrangements of panels. As far as appearance is concerned it is often impossible to tell from inspection of tho finished balloon what system has been used. The term gore, when used in connection with the conical ring system, simply means the series of panels bounded by two adjacent vertical meridian seams.

The proper valve and appendix diameters may be obtained from the Safety Code. For a balloon which is to be frequently used, it will be found an economy to have an over-size valve as this reduces the number of rip-panel landings.

The appendix length is mainly governed by practical considerations. Witb the modern "close coupled" balloon it is usually limited to about four times the diameter of the appendix opening. The present preference, especially in rubberized balloons, is to do without the old "appendix-ring" by attaching the bridle cords permanently to the bottom of the appendix. The only disadvantage of this is that the 2-ply fabric and connection patches, required for strength, introduce too much stiffness and weight for the appendix to close tightly when the balloon is descending. It is, therefore recommended that an extra single-ply sleeve is attached, inside the main appendix which is free to function as a check valve. This sleeve should have a length at least equal to its diameter.

The valve-cord may be brought down through the appendix, but the ripcord must have a separate sleeve, made to be tight at least when the balloon descends.

The rip-panel is commonly made in the form of a narrow slit with a length equal to about one-fifth of the balloon's circumference. A much better design with a more positive discharge area and minimum length of seam for reinsertion is the triangular or quadrilateral type as used in the A.D.C. balloons. (For details see "Accessories and Controls").

The number of rings is: $n^1 = \dfrac{H}{1.15}$

The number of rectangles is: $n^2 = \dfrac{2\pi R}{1.15}$ 1m.10 will be obtained, for instance.

n^1 and n^2 generally not being whole numbers.

Let us now take the whole numbers n^1 and n^2 immediately above. They will represent the numbers of bands of fabric used. The width of these bands will be obtained by dividing H by n^1 and $2\pi R$ by n^2.

To sew the seams an extra width of 3 cm. will be taken for the overlapping of the fabrics, which will then be cut out to 1m.13.

CASE OF THE SPHERICAL BALLOON—DRAWING OF THE SEGMENT.

The spherical cap not being extendable, it is divided into a certain number of quarters or segments in sufficient number to give it

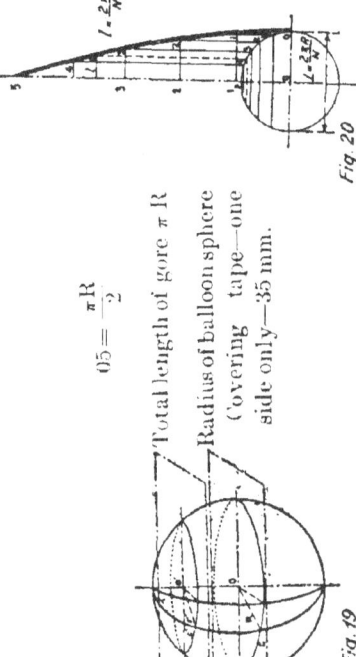

Fig. 19

Total length of gore π R

Radius of balloon sphere
(covering tape—one side only—35 mm.)

$O5 = \dfrac{\pi R}{2}$

Fig. 20

a shape resembling the sphere, without increasing the number of seams, and consequently the weight, in too large proportions.

The dimensions of the cap are determined as follows:

Let us take as base of the segment the width L of the fabric. Their number will be, for a radius R of the sphere:

$$n = \frac{2}{L}\,\pi\,R$$

It is necessary, the value of L is modified so that the number n may be whole and even.

The length of the extended axle of the segment is π R.

In order that the segment may be of proper shape, it is necessary to determine the width at points sufficiently close to each other.

Let us take, for instance, the point c, which is found again on the sphere at the point c'.

The width of the segment at that point is evidently $\dfrac{2\pi r}{n}$

For the other points the following would also be found:

$$\frac{2\pi r'}{n} \quad \frac{2\pi r''}{n} \text{ etc.}$$

The different widths are therefore proportional to the radius of R of the balloon and to the successive radii r, r', r'', etc.

All the circles being alike, let us trace on the width L of the segment a circle of L in diameter and let us divide its circumference into as many equal portions as we have taken on the segment.

Let us draw the chords, passing through these points of division which are parallel to the horizontal diameter. The lengths of these different chords give the corresponding widths of the segment.

Once the segment is traced, it is necessary to leave on one side an overlapping band of 30 m/m in order to assemble and sew together the different segments.

The pieces once cut out, the balloon is built by pasting together the different elements, which are afterwards sewn.

SEAMS.

There are several types of seams, made either by hand or by machine.

STITCHES MADE BY HAND.

1. **Basting stitch.**—The basting stitch presents no solidity whatever; it is used solely for basting. (Fig. 21.)

2. **Backstitch.**—The backstitch is made by passing the thread again in the hole of the preceding stitch.

It is the only stitch to be used for handmade seams. Seams made this way are very tight on the top and present, underneath, overlapped threads. They are very strong (five stitches per centimeter). (Fig. 22.)

3. **Saddler stitch.**—Handmade with two crossing threads, each one forming a very tight basting stitch. (Fig. 23.)

4. **Junction stitch.**—The junction stitch serves for temporary and rapid repairs. It is used for bringing together the edges of two fabrics after a tear, for instance. (Fig. 24.)

NET TENSIONS.

Contents of balloon.	Normal load.	Necessary increase for diamond nets.	Load for which net must be tested.	Breaking stress for net.	Load for which suspending net and car lines must be tested.	Breaking stress for car lines and suspending net.
V.	N.	P=1·414 N.	2 P.	10 P.	2 N.	10 N.
cb. m.	kg.	kg.	kg.	kg.	kg.	kg.
100	100	141·4	283	1,414	200	1,000
150	150	212·1	424	2,121	300	1,500
200	200	282·8	566	2,828	400	2,000
250	250	353·5	707	3,535	500	2,500
300	300	424·2	848	4,242	600	3,000
350	350	492·9	989	4,929	700	3,500
400	400	565·6	1131	5,656	800	4,000
500	450	636·3	1273	6,363	900	4,500
600	500	707·0	1414	7,070	1000	5,000
700	550	777·7	1555	7,777	1100	5,500
800	600	848·4	1697	8,484	1200	6,000
900	650	919·1	1838	9,191	1300	6,500
1000	700	989·8	1980	9,898	1400	7,000
1200	750	1060·5	2121	10,605	1500	7,500
1400	800	1131·2	2262	11,312	1600	8,000
1600	850	1201·9	2404	12,019	1700	8,500
1800	900	1276·6	2545	12,766	1800	9,000
2000	950	1343·3	2687	13,433	1900	9,500
2500	1000	1414·0	2828	14,140	2000	10,000

net and supporting cords, which must have together, in each row, a breaking stress = 10 P.

If we divide the number of mesh cords, goose's neck cords, running out lines, and car lines into the figure in the fifth column, corresponding to the size of balloon employed, we obtain the necessary breaking tension of a single line (p).

The cross section (s) can be determined if we know the breaking modulus K of the corresponding material.

$$s = \frac{p}{K} \text{ in sq. mm.,}$$

whence the diameter :

$$D = 2\sqrt{\frac{p}{K\pi}} .$$

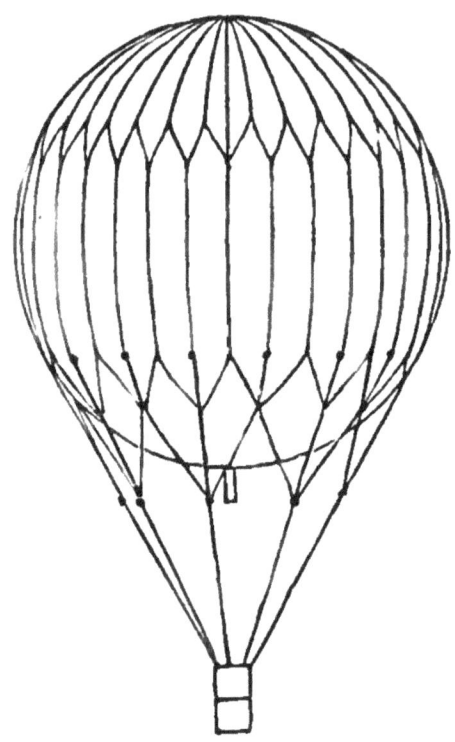

FIG. 26.—Meridian net.

Preservation of the net.—Protection against moisture and rot ; paraffin, soaking or saturating with sodium acetate. Dry the net in the sun as soon as it has become thoroughly wet.

(b) **The square net.**—Does not increase in length in the direction of the parallels, presses the covering together very strongly in this direction when wet. Not to be recommended.

(c) **The meridian net.**—According to Finsterwalder's recommendations, only to be used when fitted with step-shaped cross strings. It is very simple and light, has everywhere the same

	V(m³)	V(ft³)	Dia.(m)	Lift (kg)	Gore length (m)	Surface Area (m²)
\multicolumn{7}{c}{STANDARD Gas Balloon Volumes}
A-3	600	(21,183)	10.465	674	16.5	344
	800	(28,244)	11.519	899	18.1	417
A-4	900	(31,774)	11.980	1011	18.8	451
	1000	(35,305)	12.408	1123	19.5	484
A-5	1200	(42,367)	13.185	1348	20.7	546

Procedure. The author recommends the following adaptation of the cylindrical gore system:

1. Determine the diameter of the balloon $D = \sqrt[3]{6V}$ where V is volume.

2. Take as the minimum number of gores $n = 2\sqrt{D}$ but never less than 10.

3. Lay out carefully a quarter of a full gore to a scale not smaller than one-tenth, using the ordinates figured from the diagram of Figure. 3.

4. Beginning at the equator, by straight lines parallel to it, cut off a series of panels of the maximum fabric width obtainable, allowing for seams.

5. If the curvature at the end of a panel (side of gore) is such that its height in the center is less than 1/200 of the semi-gore width, without serious error it may be made straight instead of curved.

6. For the actual gross pattern dimensions add the width of the seam on both sides, or half the width of the seam all around.

7. If the method of setting requires the panels to be staggered, make the stagger as little as possible, about 1 in., and reverse it in the next panel of the same gore. - In this way curved horizontal seams are avoided.

The above proedure applies to a standard type of balloon using practically any kind of fabric. If 2-ply bias fabric is used it is unnecessary to alternate it, right and left hand, as is the case with airship envelopes.

SPHERICAL BALLOON
Gore Coordinates

FIG-2 = 30-PANELS

```
 1 - .05234
 2 - .10453
 3 - .13543
 4 - .20791
 5 - .25882
 6 - .30902
 7 - .35837
 8 - .40674  sines
 9 - .45399
10- .50000
11- .54464
12- .58778
13- .62932
14- .66913
```
Z -
```
15- .70711
16- .74314
17- .77715
18- .80902
19- .83867
20- .86603
21- .89101  COsines
22- .91354
23- .93358
24- .95106
25- .96598
26- .97815
27- .98769
28- .99452
29- .99863
30- 1 (equator)
```

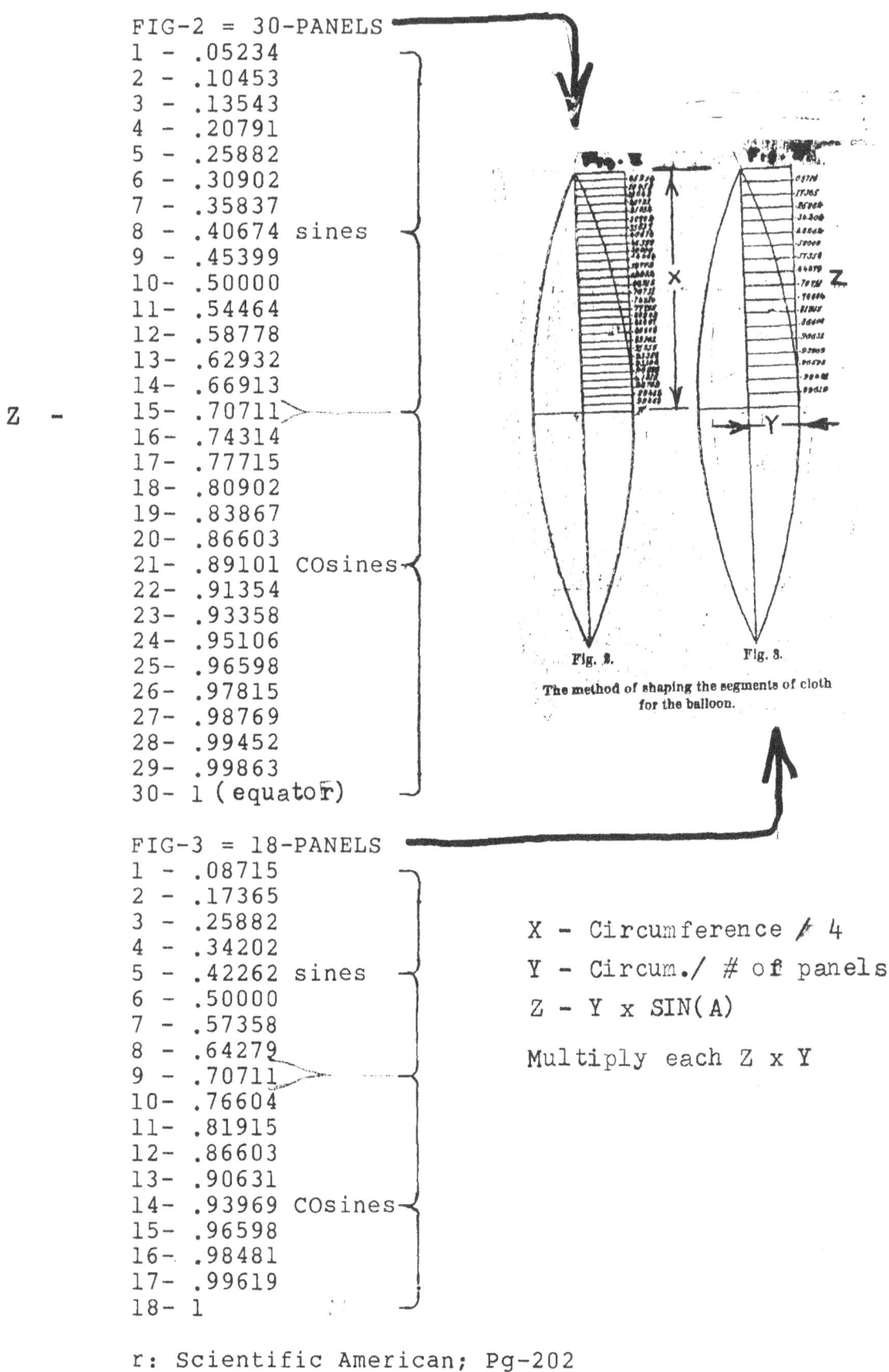

Fig. 2.   Fig. 3.

The method of shaping the segments of cloth
for the balloon.

FIG-3 = 18-PANELS
```
 1 - .08715
 2 - .17365
 3 - .25882
 4 - .34202
 5 - .42262  sines
 6 - .50000
 7 - .57358
 8 - .64279
 9 - .70711
10- .76604
11- .81915
12- .86603
13- .90631
14- .93969  COsines
15- .96598
16- .98481
17- .99619
18- 1
```

X - Circumference / 4

Y - Circum./ # of panels

Z - Y x SIN(A)

Multiply each Z x Y

r: Scientific American; Pg-202
28 Mar.1901 (SUP.vol.#65)

```
1 '*SPHERICAL BALLOON CALCS"*
2 'STORAGE CODE "SB"
3 'INPUT VM=VOL.IN CUBIC METERS
4 VM=600
5 'INPUT NG=# OF GORES
6 NG=20
7 'INPUT NI=# PATTERN INCRAM.
8 NI=30
10 REM "OUTPUT IN METRIC (M)"
11 'RM=RADIUS
12 'DM=DIAMETER
13 'CM=CIRCUMFERENCE
14 'GM=GORE LENGTH (TO APEXES)
15 'SM=SURFACE AREA IN SQ.METER
16 'LM=LIFT (HYDROGEN) IN KILOS
17 'HM=HEIGHT ON (GORE) PANEL
18 'WM=WIDTH OF (GORE) PANEL
20 REM "OUTPUT IN ENGLISH (E)"
21 'RESERVED TO LINE 29
30 REM "PROGRAM CONSTANTS"
31 K1=.6204:          ' (MATH BOOK)
32 K2=3.1416:            '(PIE)
33 K3=15.99638:       '(MATH BOOK)
34 K4=35.30554:       '(IN./METER)
35 K5=57.29578:         '(DEG/RAD)
36 K6=2.204623:       '(LBS./KILO)
37 K7=69.5:           '(H-LBS./1000)
38 K8=1.1234:         '(H-KGS./CU.M)
40 REM "B-CALCS IN ENGLISH"
42 VE=VM*K4: E1=VE^(1/3)
44 RE=E1*K1: DE=RE*2
46 CE=DE*K2: LE=VE/K7
48 SE=K2*DE*DE: GE=CE/2
50 REM "B-CALCS IN METRIC"
52 M1=VM^(1/3)
54 RM=M1*K1: DM=RM*2
56 CM=DM*K2: LM=VM*K8
58 SM=K2*DM*DM: GM=CM/2
60 REM "PRINT OUTPUT"
61 PRINT STRING$(1," ")
62 PRINT "BALLOON DIMENSIONS"
63 PRINT STRING$(1," ")
64 PRINT "VOLUME="VM"CU.METERS"
65 PRINT"DM="DM,"CM="CM
66 PRINT"GM="GM,"SM="SM
67 PRINT "LIFT="LM"KILOS"
68 PRINT STRING$(1," ")
69 PRINT "PANEL UNITS (METRIC)"
70 PRINT"(TOP) HEIGHT & WIDTH"
80 REM "PATTERN DIMENSIONS"
81 X=0
82 X=X+1: NP=NG*2
84 Y=90/NI: Z=X*Y
86 A=SIN((Z)/(K5))
88 PH=CM/4
90 HM=((PH/30)*(X))
92 M2=CM/NP
94 WM=M2*A
96 PRINT USING "##.###";X,HM,WM
98 IF X<NI GOTO 82
99 PRINT"          EQUATOR"
```

BALLOON DIMENSIONS

VOLUME= 1200 CU.METERS
DM= 13.1854675   CM= 41.4234648
GM= 20.7117324   SM= 546.187749
LIFT= 1348.08 KILOS

PANEL UNITS (METRIC)
(TOP) HEIGHT & WIDTH

1.000	0.345	0.054
2.000	0.690	0.108
3.000	1.036	0.162
4.000	1.381	0.215
5.000	1.726	0.268
6.000	2.071	0.320
7.000	2.416	0.371
8.000	2.762	0.421
9.000	3.107	0.470
10.000	3.452	0.518
11.000	3.797	0.564
12.000	4.142	0.609
13.000	4.488	0.652
14.000	4.833	0.693
15.000	5.178	0.732
16.000	5.523	0.770
17.000	5.868	0.805
18.000	6.214	0.838
19.000	6.559	0.869
20.000	6.904	0.897
21.000	7.249	0.923
22.000	7.594	0.946
23.000	7.939	0.967
24.000	8.285	0.985
25.000	8.630	1.000
26.000	8.975	1.013
27.000	9.320	1.023
28.000	9.665	1.030
29.000	10.011	1.034
30.000	10.356	1.036

EQUATOR

BALLOON DIMENSIONS

VOLUME= 600 CU.METERS
DM= 10.4653125   CM= 32.8778258
GM= 16.4389129   SM= 344.076721
LIFT= 674.04 KILOS

PANEL UNITS (METRIC)
(TOP) HEIGHT & WIDTH

1.000	0.274	0.043
2.000	0.548	0.086
3.000	0.822	0.129
4.000	1.096	0.171
5.000	1.370	0.213
6.000	1.644	0.254
7.000	1.918	0.295
8.000	2.192	0.334
9.000	2.466	0.373
10.000	2.740	0.411
11.000	3.014	0.448
12.000	3.288	0.483
13.000	3.562	0.517
14.000	3.836	0.550
15.000	4.110	0.581
16.000	4.384	0.611
17.000	4.658	0.639
18.000	4.932	0.665
19.000	5.206	0.689
20.000	5.480	0.712
21.000	5.754	0.732
22.000	6.028	0.751
23.000	6.302	0.767
24.000	6.576	0.782
25.000	6.850	0.794
26.000	7.124	0.804
27.000	7.398	0.812
28.000	7.671	0.817
29.000	7.945	0.821
30.000	8.219	0.822

EQUATOR

BALLOON DIMENSIONS

VOLUME= 800 CU.METERS
DM= 11.5185669   CM= 36.1867296
GM= 18.0933648   SM= 416.819264
LIFT= 898.72 KILOS

PANEL UNITS (METRIC)
(TOP) HEIGHT & WIDTH

1.000	0.302	0.047
2.000	0.603	0.095
3.000	0.905	0.142
4.000	1.206	0.188
5.000	1.508	0.234
6.000	1.809	0.280
7.000	2.111	0.324
8.000	2.412	0.368
9.000	2.714	0.411
10.000	3.016	0.452
11.000	3.317	0.493
12.000	3.619	0.532
13.000	3.920	0.569
14.000	4.222	0.605
15.000	4.523	0.640
16.000	4.825	0.672
17.000	5.126	0.703
18.000	5.428	0.732
19.000	5.730	0.759
20.000	6.031	0.783
21.000	6.333	0.806
22.000	6.634	0.826
23.000	6.936	0.845
24.000	7.237	0.860
25.000	7.539	0.874
26.000	7.840	0.885
27.000	8.142	0.894
28.000	8.444	0.900
29.000	8.745	0.903
30.000	9.047	0.905

EQUATOR

BALLOON DIMENSIONS

VOLUME= 900 CU.METERS
DM= 11.9797923   CM= 37.6357154
GM= 18.8178577   SM= 450.868053
LIFT= 1011.06 KILOS

PANEL UNITS (METRIC)
(TOP) HEIGHT & WIDTH

1.000	0.314	0.049
2.000	0.627	0.098
3.000	0.941	0.147
4.000	1.255	0.196
5.000	1.568	0.244
6.000	1.882	0.291
7.000	2.195	0.337
8.000	2.509	0.383
9.000	2.823	0.427
10.000	3.136	0.470
11.000	3.450	0.512
12.000	3.764	0.553
13.000	4.077	0.592
14.000	4.391	0.630
15.000	4.704	0.665
16.000	5.018	0.699
17.000	5.332	0.731
18.000	5.645	0.761
19.000	5.959	0.789
20.000	6.273	0.815
21.000	6.586	0.838
22.000	6.900	0.860
23.000	7.214	0.878
24.000	7.527	0.895
25.000	7.841	0.909
26.000	8.154	0.920
27.000	8.468	0.929
28.000	8.782	0.936
29.000	9.095	0.940
30.000	9.409	0.941

EQUATOR

BALLOON DIMENSIONS

VOLUME= 1000 CU.METERS
DM= 12.408   CM= 38.9809728
GM= 19.4904864   SM= 483.675911
LIFT= 1123.4 KILOS

PANEL UNITS (METRIC)
(TOP) HEIGHT & WIDTH

1.000	0.325	0.051
2.000	0.650	0.102
3.000	0.975	0.152
4.000	1.299	0.203
5.000	1.624	0.252
6.000	1.949	0.301
7.000	2.274	0.349
8.000	2.599	0.396
9.000	2.924	0.442
10.000	3.248	0.487
11.000	3.573	0.531
12.000	3.898	0.573
13.000	4.223	0.613
14.000	4.548	0.652
15.000	4.873	0.689
16.000	5.197	0.724
17.000	5.522	0.757
18.000	5.847	0.788
19.000	6.172	0.817
20.000	6.497	0.844
21.000	6.822	0.868
22.000	7.147	0.890
23.000	7.471	0.910
24.000	7.796	0.927
25.000	8.121	0.941
26.000	8.446	0.953
27.000	8.771	0.963
28.000	9.096	0.969
29.000	9.420	0.973
30.000	9.745	0.975

EQUATOR

APPENDIX D.

CONSTRUCTION OF BALLOON.

We have now determined[1] the length of the whole gore, and the length of thirteen divisions of each of the two semi-prolate spheroids of which it is composed, and corresponding to twenty-seven points selected along the axis of the balloon, two of which points, viz. No. 1 and No. 27, are the fore and aft extremities.

The first two columns represent the lengths or distances to be measured on the length of the gore, beginning at the fore end.

The first column being the entire distance of each successive point from the fore extremity.

The second column representing the distances of each of the same measured from the preceding point in the same direction.

The third column gives the girth of the balloon at each measured point of the length of the gore, viz. the point in the same horizontal line in the Table.

From the numbers in this column the breadth of the gore can be deduced by dividing each number by the number of gores intended to be used.

The two final columns give the elliptic coordinates of the section of the balloon at the corresponding points, supposing the figure completed.

The unit of measurement in all is the radius of the greater circle, viz. that at the minor axis of the two combined spheroids.

[1] The calculations from which this result has been derived, and which are of considerable length, are omitted.—Ed.

Table showing the dimensions of gores for the half-prolate spheroidal balloon ($a = 4b$ & $a' = 8b$) in terms of minor axis of generating ellipses taken as unity.

Points on the gore in numerical order	Dimensions of gores			Corresponding dimensions of spheroid	
	Distances along the gore measured from the fore end	Partial lengths along the gore measured to each point from the preceding one	Circular girth of the gore at each point	Ordinates or semi-diameters measured at right angles to the axis of the balloon	Horizontal measurements or *abscissæ* measured from the centre of each spheroid along the axis of the balloon
1	0·000	0·000	0·000	0·000	4·000
2	0·044	0·044	0·275	0·044	3·996
3	0·090	0·046	0·435	0·087	3·984
4	0·137	0·047	0·820	0·130	3·966
5	0·187	0·050	1·091	0·174	3·940
6	0·302	0·115	1·626	0·259	3·863
7	0·436	0·134	2·149	0·342	3·759
8	0·771	0·335	3·142	0·500	3·464
9	1·195	0·424	4·039	0·643	3·064
10	1·704	0·509	4·814	0·766	2·571
11	2·283	0·579	5·411	0·866	2·000
12	2·919	0·636	5·905	0·940	1·368
13	3·595	0·676	6·185	0·985	0·695
14	4·289	0·694	6·283	1·000	0·000
15	5·678	1·389	6·185	0·985	1·389
16	7·026	1·348	5·905	0·940	2·736
17	8·292	1·266	5·441	0·866	4·000
18	9·439	1·147	4·814	0·766	5·143
19	10·432	0·993	4·039	0·643	6·128
20	11·245	0·813	3·142	0·500	6·928
21	11·857	0·611	2·149	0·342	7·518
22	12·082	0·225	1·626	0·259	7·727
23	12·257	0·175	1·091	0·174	7·880
24	12·325	0·068	0·820	0·130	7·932
25	12·383	0·058	0·435	0·087	7·969
26	12·434	0·051	0·275	0·044	7·993
27	12·479	0·045	0·000	0·000	8·000

ascensive powers of balloons, computed at one ounce of power for every foot capacity, so that, if carburetted hydrogen or coal gas is used, allowance must be made accordingly. The minute fractions of feet and ounces will not be noticed, as they are of no consequence in the mere practice of the art. The surfaces may be converted into square yards by dividing them by 9.

Feet diameter.	Surfaces in square feet.	Capacities in cubic feet.	Pounds ascensive power.
1	$3\frac{1}{16}$	$0\frac{1}{2}$	$0\frac{1}{32}$
2	$12\frac{1}{2}$	4	$0\frac{1}{4}$
3	28	14	1*
4	50	33	2
5	78	65	4
6	113	113	7
7	154	179	11
8	201	268	17
9	254	381	24
10	314	523	33
11	380	697	44
12	452	905	57
13	531	1,150	72
14	616	1,437	90
15	707	1,767	110
16	804	2,145	134
17	908	2,572	161
18	1,018	3,054	191
19	1,134	3,591	224
20	1,257	4,189	261
21	1,385	4,849	302
22	1,520	5,575	348
23	1,662	6,371	398
24	1,810	7,238	452
25	1,963	8,181	511
26	2,124	9,203	575
27	2,290	10,306	644
28	2,463	11,494	718
29	2,642	12,770	798
30	2,827	14,137	884
31	3,019	15,598	975
32	3,217	17,157	1,072
33	3,421	18,817	1,176
34	3,632	20,580	1,286
35	3,848	22,449	1,403
36	4,072	24,429	1,527
37	4,301	26,522	1,658
38	4,536	28,731	1,796
39	4,778	31,060	1,942
40	5,026	33,510	2,094
45	6,362	47,713	2,982
50	7,854	65,450	4,091
55	9,503	87,114	5,445
60	11,310	113,098	7,069
65	13,273	143,794	8,987
70	15,394	179,595	11,225
75	17,671	220,804	13,800
80	20,106	268,083	16,755
85	22,698	321,556	20,097
90	25,547	381,704	23,856
95	28,353	448,922	28,058
100	31,416	523,599	32,725
200	125,664	4,188,792	261,800
400	502,656	33,510,336	2,094,400
800	2,010,624	268,082,688	16,755,200

Radius	Volume	(964 #/1000) LIFT Kg. based on M^3	(60# /1000) lbs. based on $Ft.^3$
1	4.1888	4.0384	.2513
2	33.5104	32.314	2.0106
3	113.098	106.457	6.7858
4	268.083	258.509	16.0849
5	523.600	504.907	31.4160
6	904.781	872.480	54.2868
7	1436.76	1385.467	86.2056
8	2144.67	2068.105	128.680
9	3053.64	2944.625	183.218
10	4188.89	4039.346	251.333
11	5575.29	5376.252	334.517
12	7238.25	6979.844	434.295
13	9202.79	8874.250	552.130
14	11494.1	11083.760	689.646
15	14137.2	13632.501	848.232
16	17157.3	16544.784	1029.438
17	21790.1	21012.193	1307.406
18	24429.1	23556.981	1465.746
19	28731.0	27705.303	1723.860
20	33510.4	32314.078	2010.624
21	38792.5	38371.907	2327.550
22	44602.3	43009.997	2676.138
23	50965.1	49145.645	3057.906
24	57906.0	55838.755	3474.360
25	65450.0	63113.435	3927.000
26	73622.3	70993.983	4417.338
27	82448.2	79504.799	4946.892
28	91952.5	88669.795	5517.150
29	110216.0	98512.888	6129.600
30	113098.	109060.401	6785.880

NOTE: The lift calculations are for standard temperature at sea level; they are basically good for hydrogen or helium, however, lift is greatly affected by the purity of the gas and variations in density altitude.

* Keep units in same values; if radius is in meters then volume will be in cubic meters and lift in kilograms.

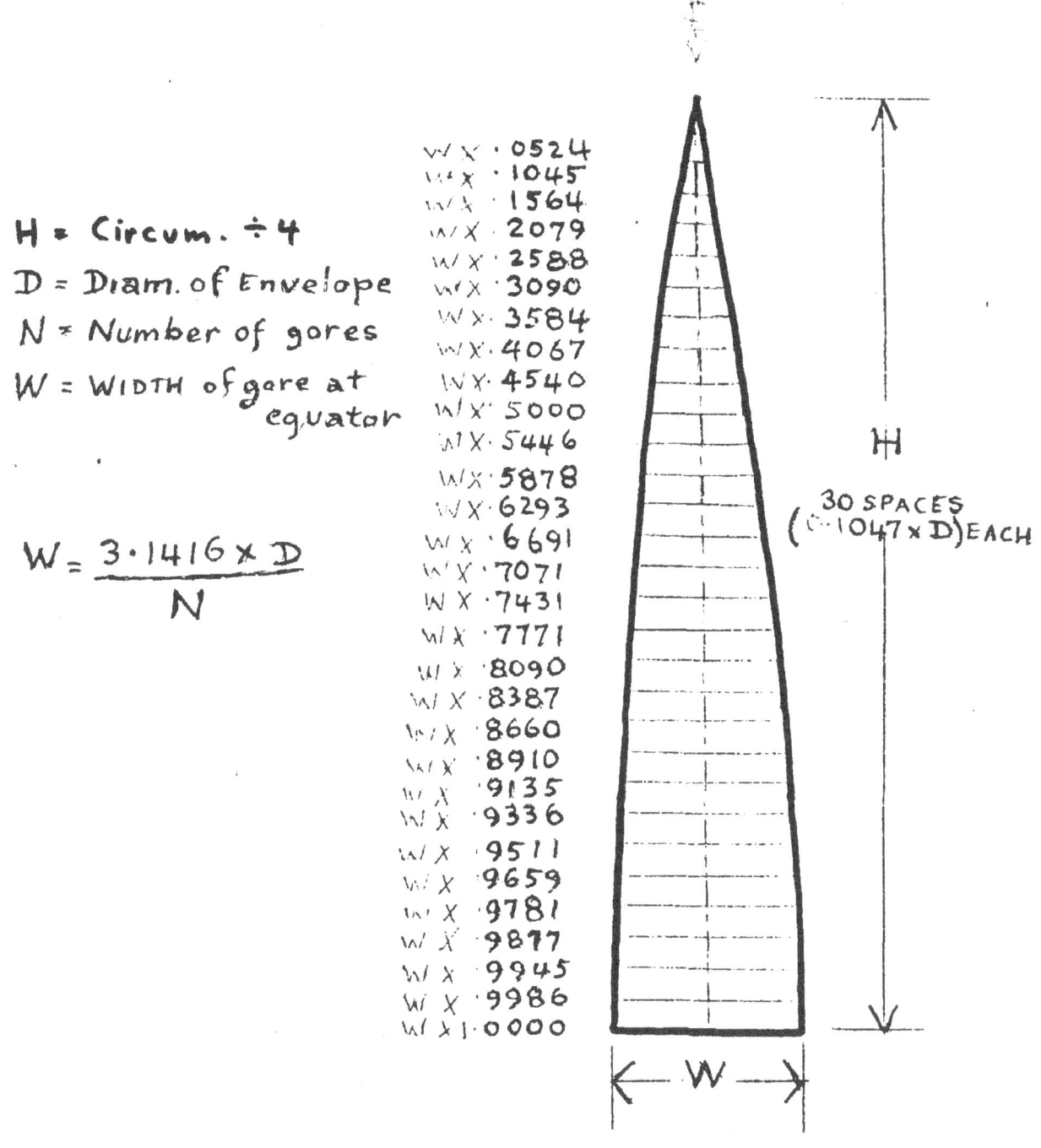

$H = $ Circum. $\div 4$

$D = $ Diam. of Envelope

$N = $ Number of gores

$W = $ WIDTH of gore at equator

$$W = \frac{3 \cdot 1416 \times D}{N}$$

W X ·0524
W X ·1045
W X ·1564
W X ·2079
W X ·2588
W X ·3090
W X ·3584
W X ·4067
W X ·4540
W X ·5000
W X ·5446
W X ·5878
W X ·6293
W X ·6691
W X ·7071
W X ·7431
W X ·7771
W X ·8090
W X ·8387
W X ·8660
W X ·8910
W X ·9135
W X ·9336
W X ·9511
W X ·9659
W X ·9781
W X ·9877
W X ·9945
W X ·9986
W X 1·0000

H

30 SPACES
(·1047 x D) EACH

W

MATHMATICAL CURVE CALCS.

ENVELOPE GORE FOR A SPHERICAL BALLOON

Net Design

The net is made of horizontal rows of diamond-shaped meshes, all meshes in the same ring being of equal size. The top of the net is terminated in a ring, preferably of coiled w1re cable, of a size to fit around the valve. The lower end of the net connects thrcugh bridles (crow's-feet) with the footropes.

The size of the rope required for tho crow's-foot and the net is determined by

$$S_1 = F \frac{W}{2n \cos a \cos \frac{B}{2}}$$

S_1 = breaking strength of the rope in lbs

n = number of meshes

B = mesh angle at top or bottom of equator meshes (usually 53°)

The ring at the top should have a strength (one side):

$$S = \frac{F}{2} n S_1$$

In nets for the larger balloons it is possible to reduce the strength gradually above the equator, until at the valve ring othe strength of the cordage need only be about two-thirds of that at the equator and below it. At the same timeo it is well to make the extreme bottom mesh about 25 per cent stronger to take up the extra wear due to sandbag hooks and crows-feet. It is sometimes cnsidered an object to have the net strong enough for climbing on it to examine the valve. This usually can be attained by keeping down the number of meshes, provided the dimensions between the knots doe not exceed 2 to 3 ft. The higher the fabric factor of safety, the larger the net meshes that can be allowed.

Inflation net. In netless balloons it is necessary to use a net for the inflation until the balloon is sufficiently full to allow the sandbags to be attached to the suspension ropes.

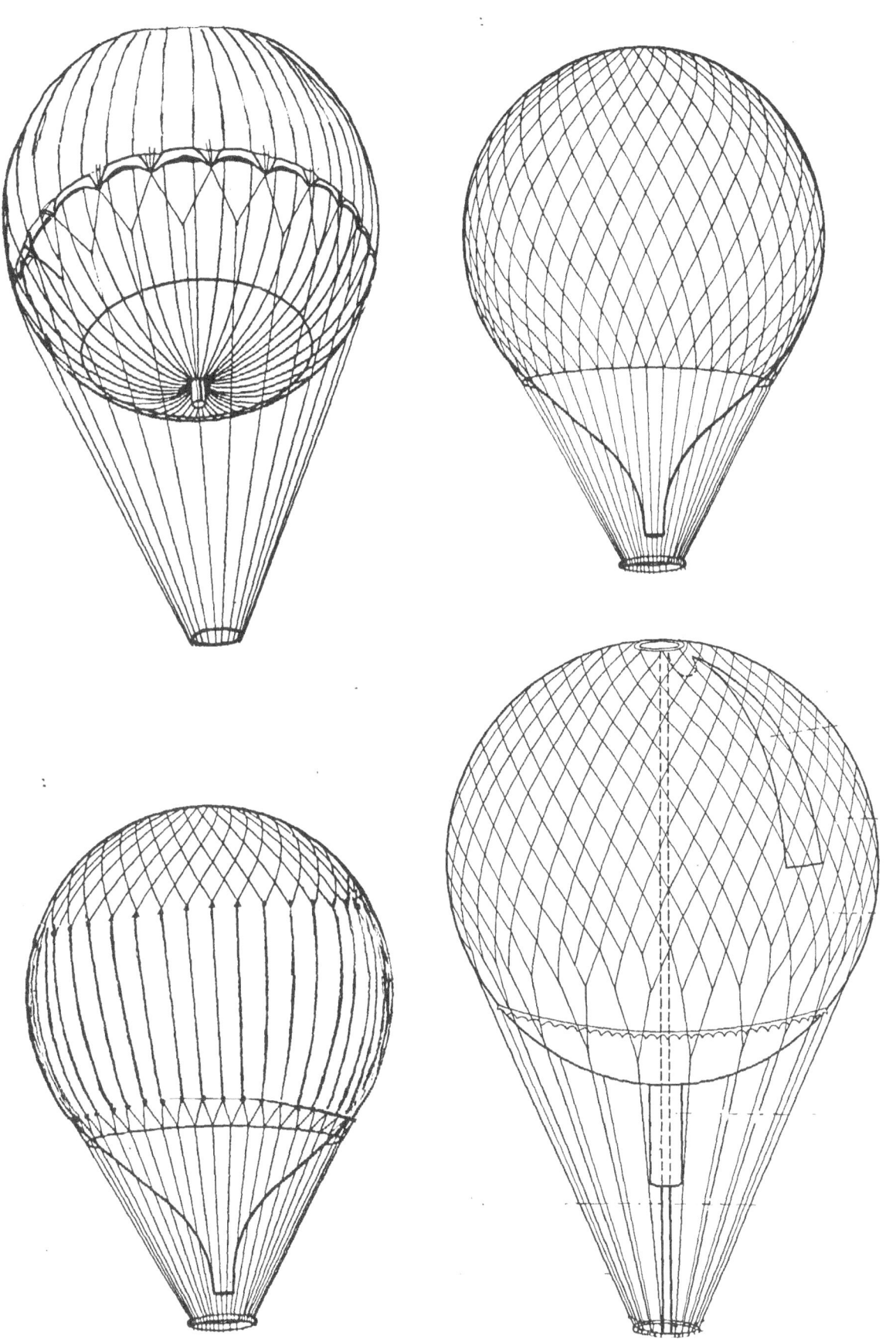

Before You Begin

Net-making is a systematic method of knotting, in which you repeat the same hand-tied knot row after row to create a fabric from one strand or cord. In this, netting is somewhat similar to crocheting or knitting in that you work back into the previous row to form a continuous whole.

Though perhaps most often associated with fishing, netting is certainly not limited to fish nets. Because of its strength and lightness, net is a natural choice for carriers and containers and can also be used to make many other useful and decorative items. With the widespread interest in crafts, soft sculpture and fabric constructions, the creative possibilities based on the net-making technique are innumerable. The aim of this book is to acquaint you with some of these possibilities.

You achieve variety in netting depending on the cords you use, their size and on the size of the mesh (the loop or space between any two knots). By increasing and decreasing, seaming and manipulating shapes, you can use netting to create a variety of forms. By combining netting with other knotted techniques, such as macramé, you may discover entirely new creative directions.

Perhaps one reason net-making is not thought of immediately as a decorative art is the lack of color in most examples. You can easily remedy this by using the colorful macramé cords available in craft and hobby shops or by using cotton seine twine which you can dye with cold-water or washing-machine dyes.

If you do want to dye your own cord, you must first decide when to do the dyeing. If you dye the cord before you make your net, you may have short pieces of odd-colored cord left over. When you use dyed cord for tassels later, you very seldom get good results. The dye has not penetrated well so you get a striped effect. Of course, not every project has tassels, and it is often advantageous to dye various sizes of cord at one time.

You may also dye the piece after it is finished, but note that it may shrink considerably and if the knots ever loosen, you will see the original color, where the dye did not penetrate. Tassels, though, will be dyed all the way through.

Whenever you decide to dye, first wash the cord or finished net in quite hot water, using detergent and, if you like, fabric softener. Rinse extremely well and follow the directions of the manufacturer of the commercial dyes.

Equipment and Tools

Cord or Twine

A few twines are made specifically for hand-netting. Bonded nylon is one. Normally, nylon is slippery and does not easily hold a knot, but the bonding process helps. Cotton seine twine is most useful for net-making, but you may want lighter weights too. Finding sizes 18 and 21 is easy, but finding smaller sizes may present a problem. You may find #12 cotton seine twine in hardware stores, but you might have to look for #9; #6 is made, but mostly for commercial users.

4

You may also net with various types of craft cords and yarns. Look for marine suppliers, hardware stores, craft shops and upholstery and weaving supply houses who carry suitable cords.

Another thing to consider in your twine selection is, of course, use. Salt water is harder on nets than fresh water, and direct sunshine is worst of all. Washing off slime as quickly as possible helps preserve the nets. You might also want to consider using some of the commercial preservatives recommended for cotton canvas for outdoor accessories.

Shuttle and Gauge

Though it is possible to net without a shuttle and gauge, it is easier and more efficient to use these tools (see Illus. 1). The shuttle is the wooden, plastic or even bone holder of twine or cord which you use to carry the twine while making the netting knot. The width of the gauge determines the size of the mesh. Using a gauge ensures that the mesh will be uniform and a specific size.

Gauges actually measure just under the size you think they are. That is, a $1''$ gauge is really about $\frac{15}{16}''$ wide. When you are first learning to net, try out a few different sizes, one size to a row, and measure the results. When you become more proficient, you can try double wrapping a gauge to make a mesh loop twice as big as the one you expect from that gauge.

If you cannot find a shuttle at your local crafts supplier, through fishermen's suppliers in your area, or by mail order, you can easily make your own. In fact, some net-makers prefer to make their own tools.

To make a shuttle, choose a hardwood that you

Illus. 1. Wooden shuttle (top) and gauge.

can polish reasonably well, so that the cord does not catch on the shuttle. About $\frac{1}{2}''$ thick is the thickest a shuttle can be and still handle easily; $\frac{3}{16}''$ thick is better and $\frac{1}{8}''$-thick boxwood or fine-grained straight maple is ideal. The shuttle and gauges shown in this book are walnut and were made with simple hand and power tools. You can use hand tools—in fact, a jackknife plus a bit of sandpaper are enough—but a power jig saw is faster. The pattern in Illus. 2A and 2B is for the most common shuttle shape, though by no means the only shape.

To make a shuttle like the one shown in Illus. 2A and 2B, trace the parts of the pattern. It is sized for use with gauges $1''$ and larger. If you want to use smaller gauges, reduce the size of the shuttle which should always be somewhat narrower than the gauge you are using. The pattern is drawn on a grid so that you can reduce or enlarge it easily.

Diamond Mesh Carrier

Materials:
#36 cotton seine twine
3" gauge

A rectangular sample of diamond meshing is an appropriate first project to acquaint you with the basic net-making tools and techniques. Finding something useful to do with such a rectangle is not hard at all. For example, you can put the piece down flat and pile it full of clothes to take to a rummage sale. Or, you can tie through the mesh and you will have an easily handled bundle. If you put down your sample in the woods and other firewood, you can tie it and tote it back to the campsite. An improvement is a double selvage edge (see instructions on page 12) if you know you are always going to hook or tie down the net through the edges. For your first project, why not make a rectangular car-top carrier, as shown in Illus. 3? Make a net no larger than 3 feet by 5 feet.

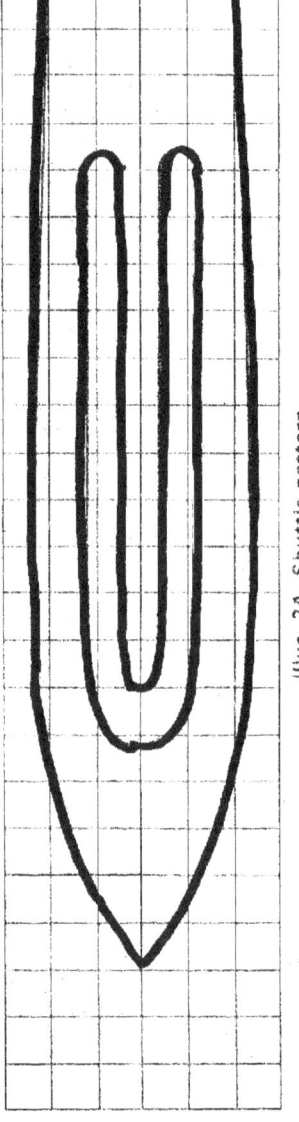

Illus. 2A. Shuttle pattern.

The length of a shuttle depends on two things: maintaining reasonable proportions so it handles well and the size cord you are using. Heavy cord wound many times on a short shuttle is not going to go through the meshes as easily as the same amount of cord wound on a longer shuttle. The pattern given produces a shuttle 10½" long that will carry about 15 yards of #48 twine or 12 yards of #36.

Other Equipment

In addition, you need a work surface into which you can drive a few nails. This can be an old wooden desk or simply be a piece of 1" × 2" wood, and a C-clamp to clamp the board onto a work table. The exact dimensions of which you clamp such a work board depends entirely on which angle you feel most comfortable with. This can mean flat on the work table, at a slightly raised angle, or even up on a wall, perpendicular to the work table. Try a few work positions to see which you prefer.

You may also use a wooden dowel, instead of a work board, which you attach to the edge of your work table with two C-clamps.

You also need scissors, occasionally a raffia needle, a good light and a comfortable chair.

Filling the Shuttle

Either use the shuttle you just made (page 6) or select one slightly smaller than the width of the gauge you will be using most. Filling the shuttle properly is the first step.

If you are using the style shuttle in Illus. 2A and 2B, hold it in your hand. Hold the free end of twine under your thumb. Bring the twine up around the tongue and down under the heel, holding down the free end as you go (see Illus. 4, left). Turn the shuttle over in your hand and carry the twine up, around the tongue, and back

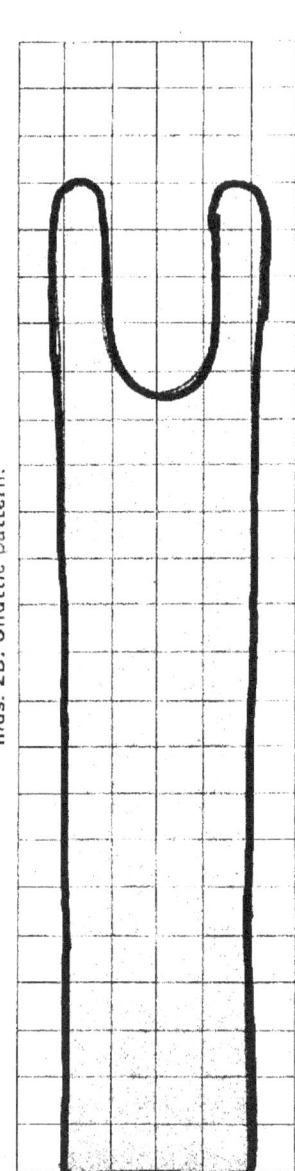

Illus. 2B. Shuttle pattern.

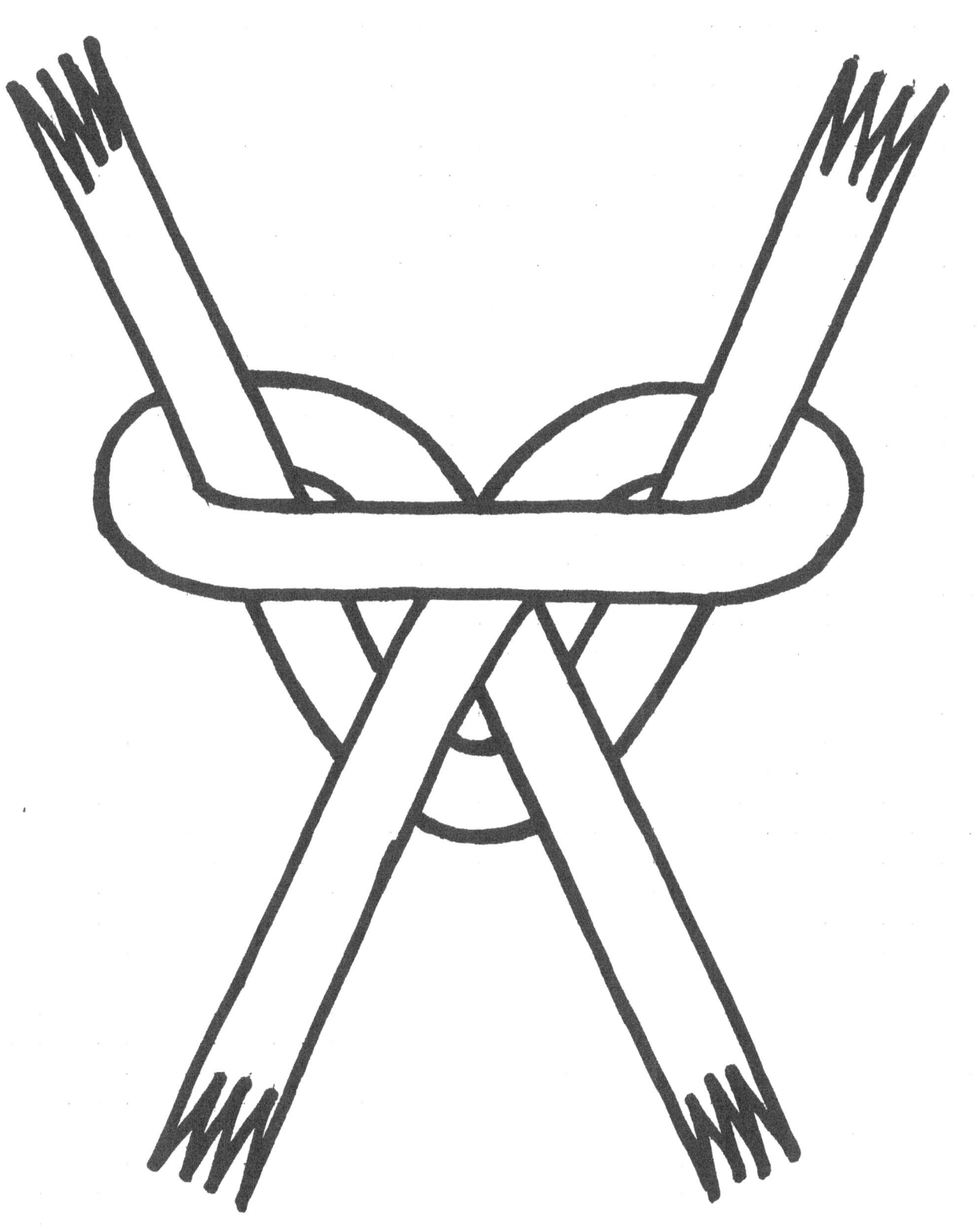

NET KNOT

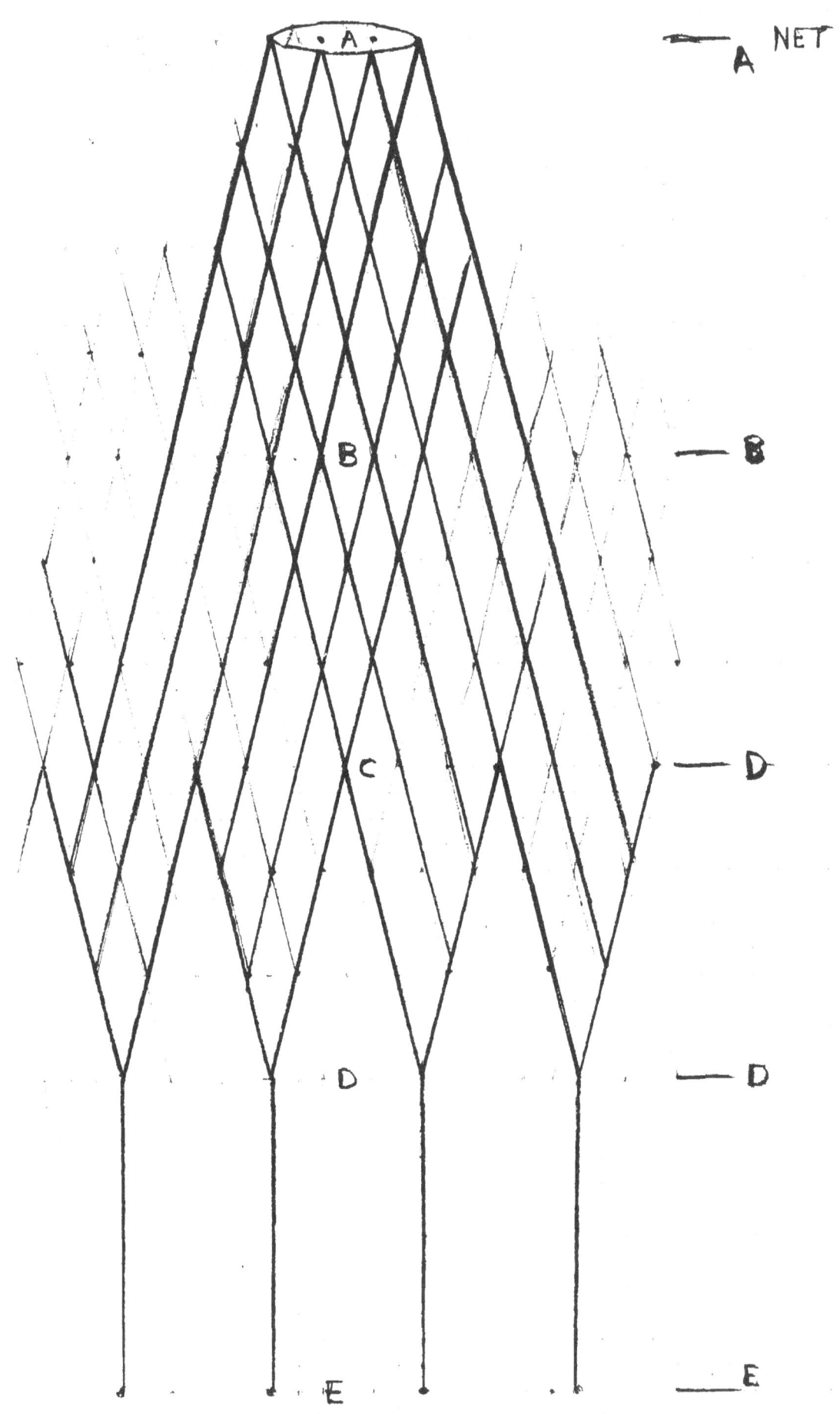

NET

A

B

D

D

E

44

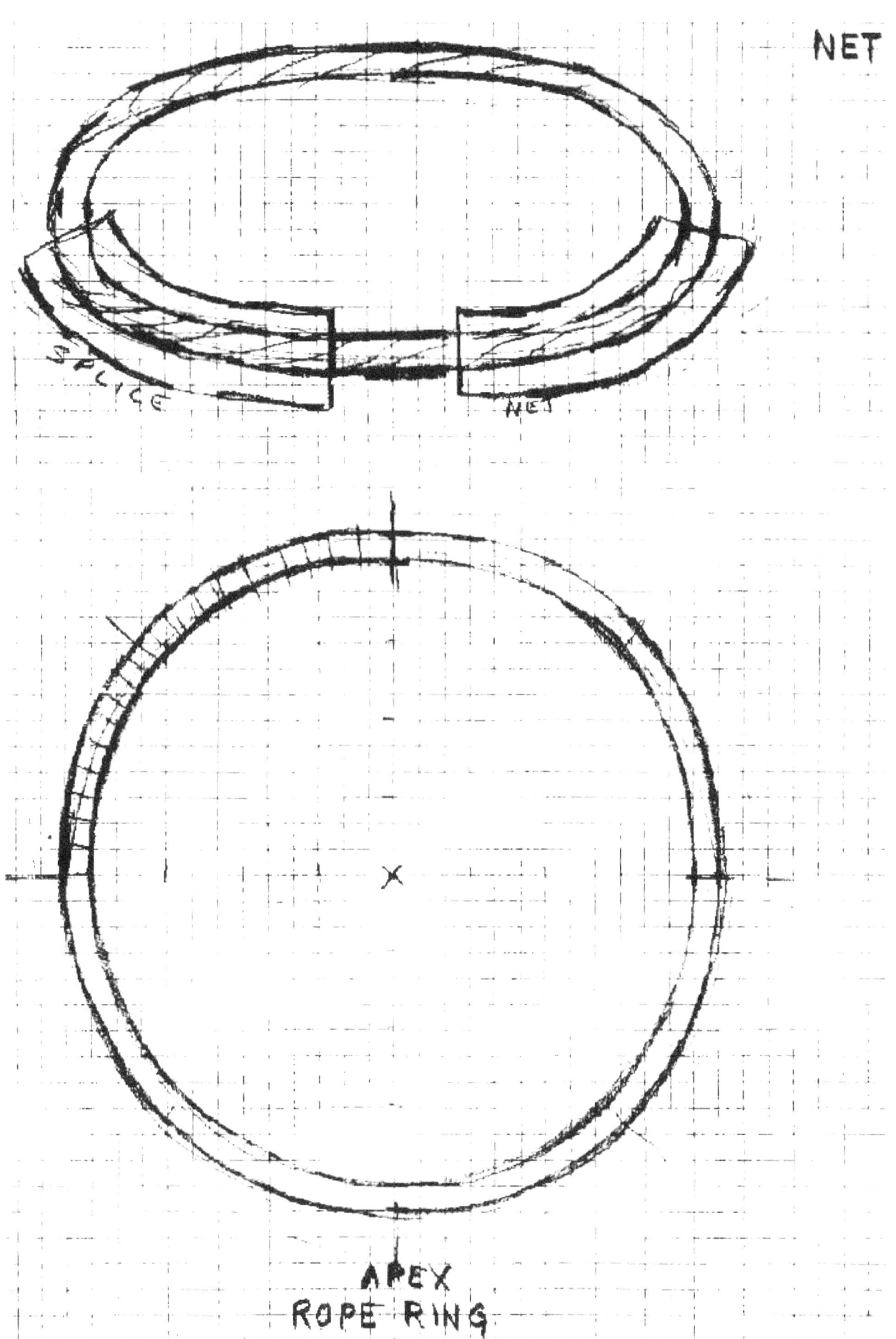

SPLICE

NET

APEX
ROPE RING

45

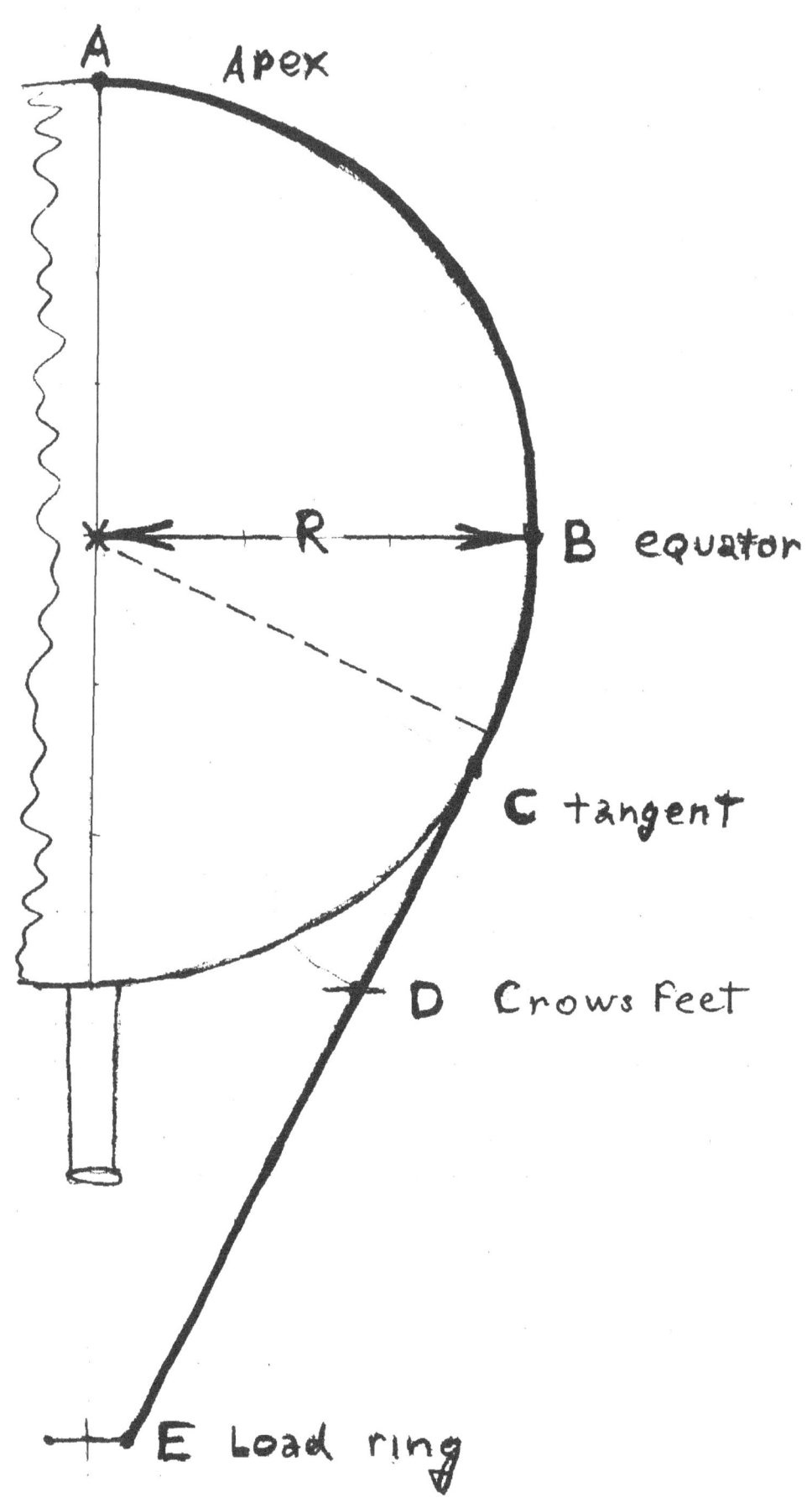

A Apex

R

B equator

C tangent

D Crows Feet

E Load ring

46

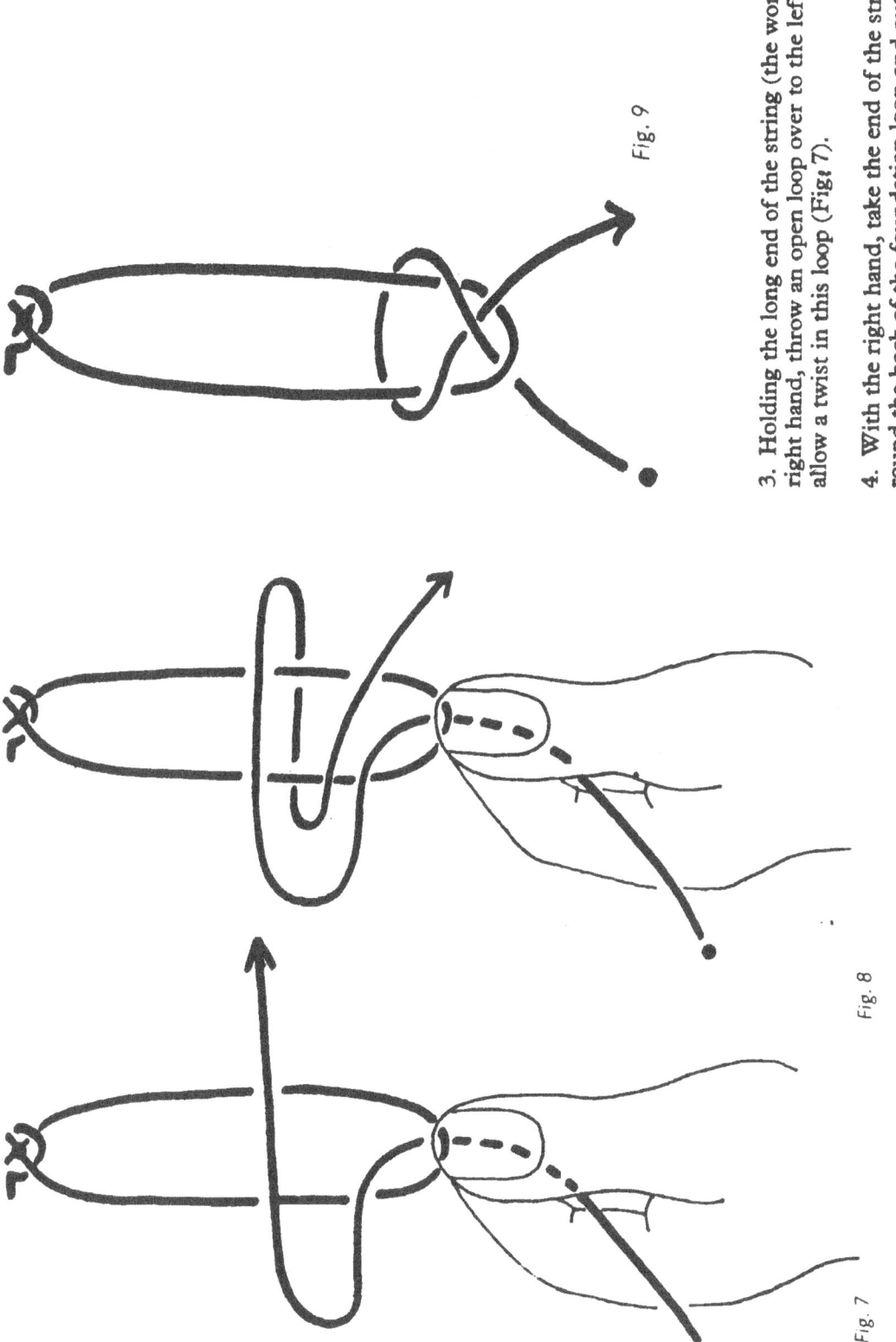

Fig. 9

Fig. 8

Fig. 7

3. Holding the long end of the string (the working end) in the right hand, throw an open loop over to the left. Take care not to allow a twist in this loop (Fig. 7).

4. With the right hand, take the end of the string to the right, round the back of the foundation loop and out to the front through the thrown loop (Figs. 8 & 9).

5. Keeping the left thumb and finger firmly in position until the last moment, pull the knot firm, but not tight at this stage.

20|21

The third and subsequent rows

1. Remove the mesh stick only at the end of each row and continue meshing successive rows, always working from left to right. The processes fall into a natural rhythm and it is useful, even desirable, to check your work at the end of each row. With children the young beginner is glad to offer the work for inspection at such intervals, especially as with increasing skill the intervals are manifestly becoming shorter each time.

2. Stop when you feel you have had sufficient practice and are ready to apply the processes to making the first article, the football or basketball carrier.

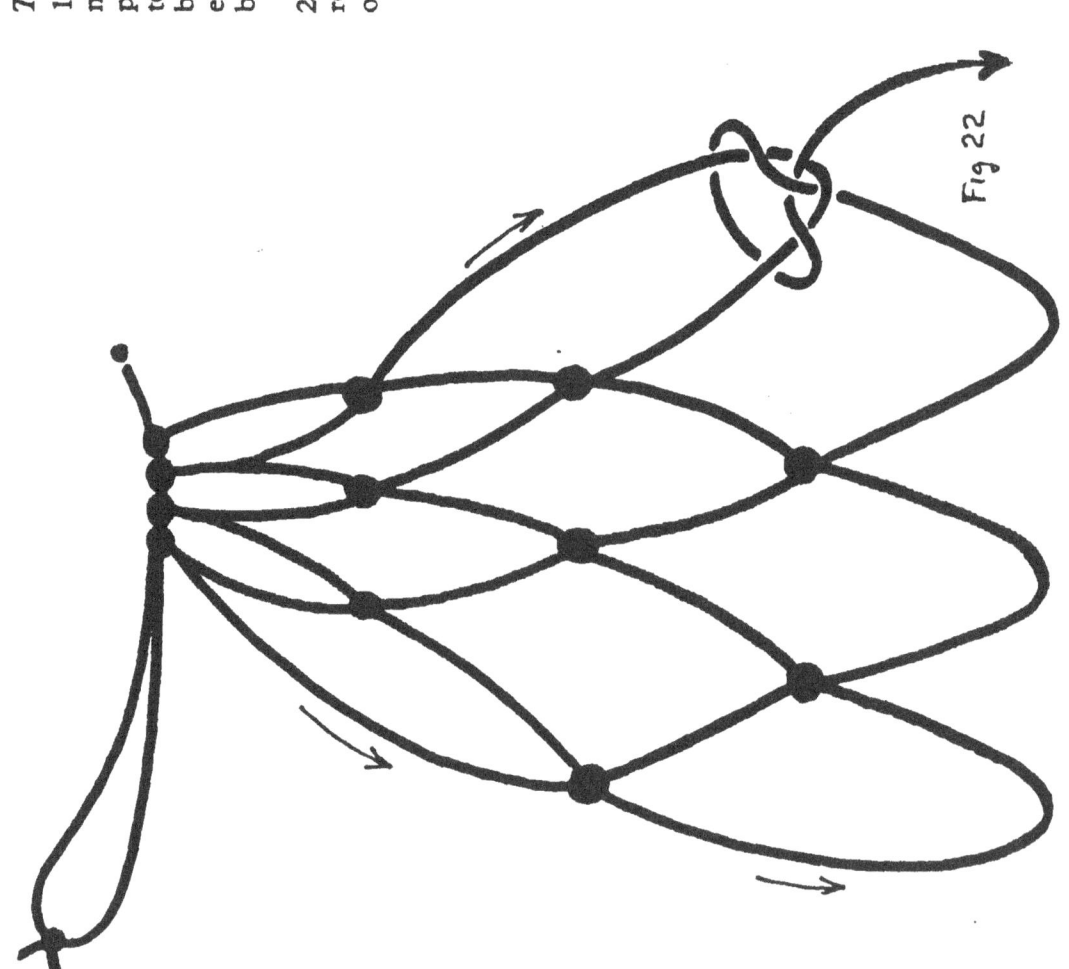

Fig 22

48

10. Repeat 10 times to make 11 loops. You should count the loops at the bottom of the mesh stick and not at the top where there will be an extra knot (Fig. 28).

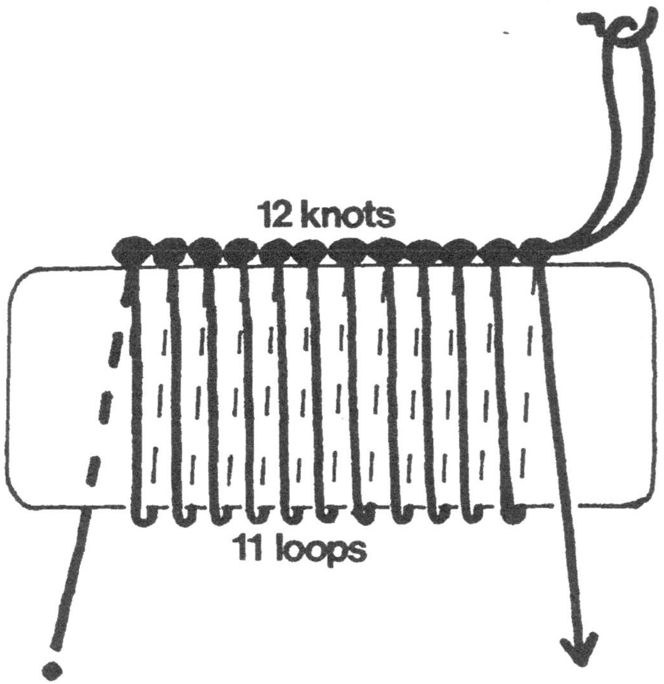

Fig. 28

12 knots

11 loops

Fig. 29

a

Fig. 30

b

11. Remove the mesh stick and tuck the tail through what is left of the foundation loop. Taking care not to withdraw the tail, pull at the point marked *a* and reduce the foundation loop to nothing (Fig. 29). This may need a fairly strong pull, depending on the tightness of your knots, but it will not come apart. Take care however not to jerk; a steady pull is needed. Do not be alarmed when a certain amount of twist forms in the loop as it is reduced. You can stop any number of times to ease it away.

12. By pulling now on the tail itself, marked *b*, draw the whole piece into its grommet shape (Fig. 30). This must be made secure by tying the tail and the working string in a good sound reef knot, close up to the other knots which form the grommet.

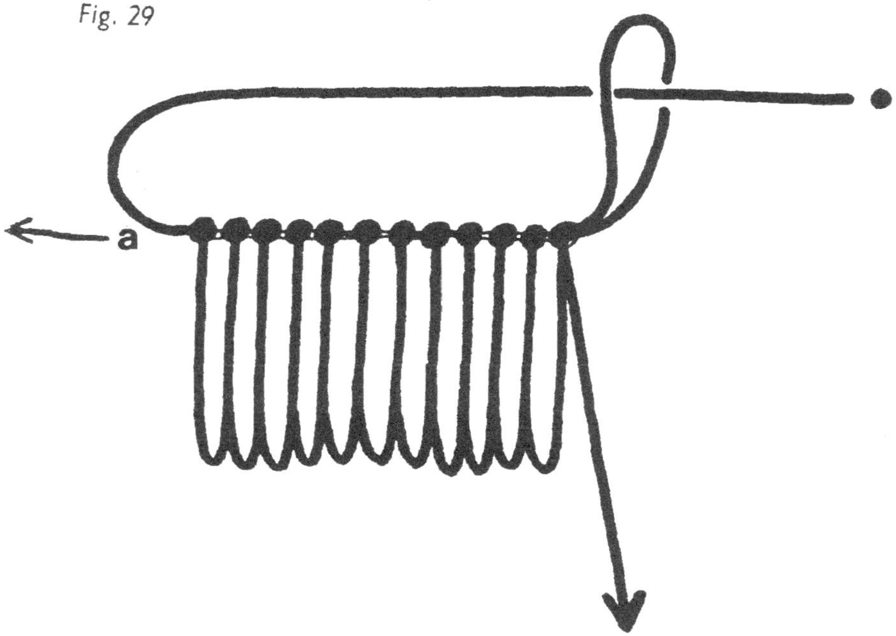

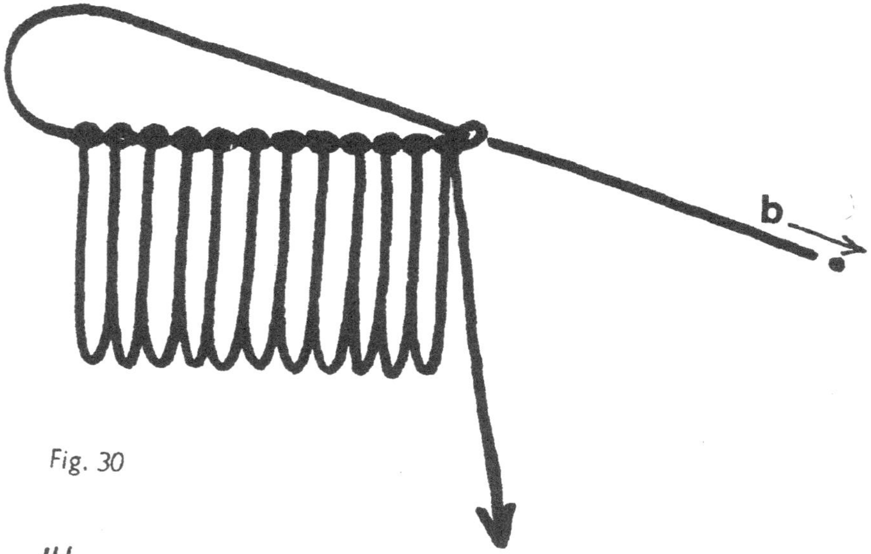

The first row

1. Mesh round in the usual way *but remove the mesh stick after every three meshes.* Do not try to force all the meshes of the complete circle on the mesh stick at once.

Take care to mesh each loop in proper sequence and to take any twist from the loops. It may help you as a beginner to have an assistant as a second pair of hands to sort the loops. With practice however you will begin to use a second finger of the left hand to do this for yourself.

2. After meshing the 11th loop, the tail and the working string must be measured with the neighbouring mesh and tied in a double overhand knot (Fig. 34).

Fig. 34

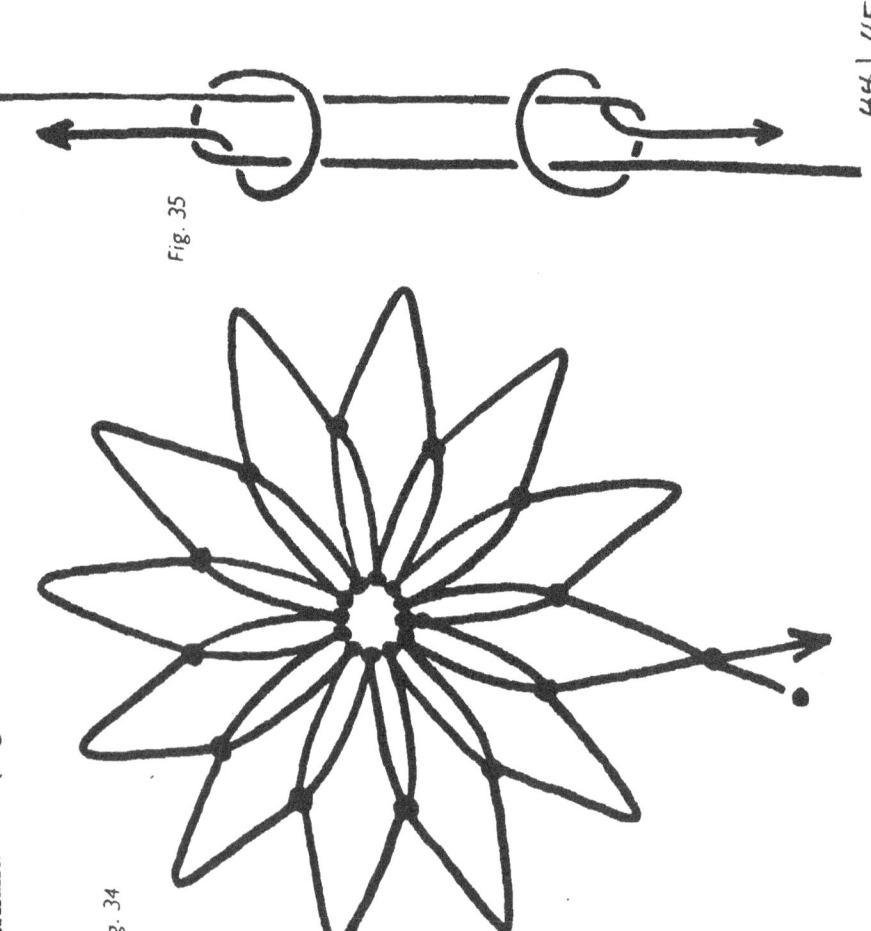

The second and subsequent rows

Continue meshing round, removing the mesh stick after every 3 meshes, and at the end of each row tying the tail and the working string in a double overhand knot.

As a check, count the meshes at the end of each round. If you have missed any meshes it will not need much taking back. There should of course be twelve meshes, including the last one made with the double overhand knot.

Note on joining on a new length of string

After about 3 rows you will have used most of the string on the needle. To join on a new length of string, tie it-on with any suitable knot, e.g. a reef knot. More sophisticated methods should be deferred until you are more accomplished.

Meanwhile a perfectly sound method, not too difficult and which appeals to children and adults, is the Fisherman's Knot (Fig. 35). This is made by making a simple overhand knot in each end round the other string. By pulling on the short ends you separate the knots and by pulling on the strings outside the knots you draw the two knots together into a firm Fisherman's Knot.

Fig. 35

Alternatively you can use the same basic netting knot by turning back the last of the working string and making a netting knot in it with the new string (Fig. 36). This is in fact a Weaver's Knot and makes a sound joint. (For a speedier method of making this knot see Chapter 5, Now you can net.)

Finishing off

1. Complete 7 rows. (Count 7 rows of knots, excluding those of the grommet.) Tie off the last double overhand knot very firmly and cut off the ends neatly and fairly short.

2. Remove the toggle.

3. Lace the top round of meshes with the thick string for a draw string. A popular finish is made by tying the draw string in a double overhand knot, leaving both ends a good inch long (Fig. 37). Strand the two ends and separate the fibers in each strand with an ordinary hair comb. This makes a convincing tassel.

44 | 45

GAS

Generating Hydrogen Gas

I. $2 HCl + Fe = H_2 + FeCl_2$

1. Use a lead-lined container

2. Acid must be oil free

3. Cool the gas with a water spray

4. Dry the gas with calcium chloride

   or caustic soda

   or un-slaked lime

II. $H_2SO_4 + Fe = H_2 + FeCl_2$

1. Dilute 98% acid with 6 gallons of water to 1 gallon of acid

2. Add acid to water (never the reverse)

Accessories and Controls[6]

The basket varies in size according to the number of passengers and other requirements. For short trips of a couple of hours duration 3 sq. ft. of floor area per passenger is sufficient, although rather crowded. For longer trips and for racing, more than double this area may be needed. In any case the minimum size of a basket - even if only one person is carried - should not be less than 30 in. in width and 36 in. in lcngth in order to afford sufficient protection when landing in a strong wind.

The height varies from 30 to 42 in.

The materials used in basket construction are willow and reed. For ordinary purposes reed is generally used, being stronger and more durable. But for racing baskets where lightness is of prime importnace the ligher willow can be used to advantage, either alone or in combination with reed. The weight can be further reduced and the basket also stiffened by inserting open spaces filled with diagonal cross-lattice work. (See Plate 1, upper right)

The balloon basket is woven like an ordinary basked and reinforced in places exposed to wear. The top edge is stiffened by a forme of wood or heavy cane to withstand the strain of the basket ropes. Across the bottom are placed several hardware crosspieces (skids) which serve to protect the bottom when in contact with the ground and stiffen the bottom so as to form a level floor. (See Plate 1, upper left)

----

[6] See also Section 80 of Safety Code Appendix III.

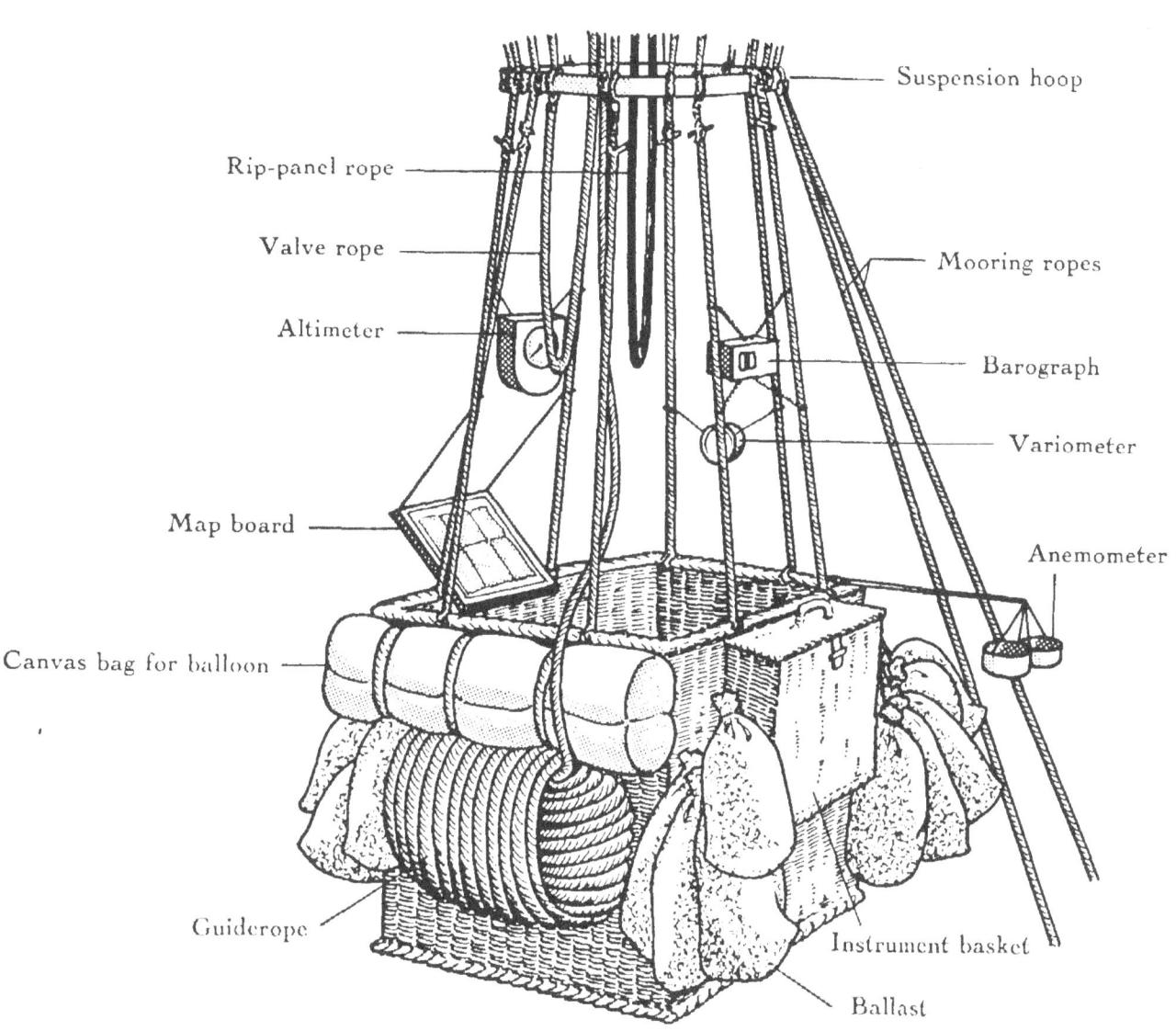

Suspension hoop

Rip-panel rope

Valve rope

Altimeter

Mooring ropes

Barograph

Variometer

Map board

Anemometer

Canvas bag for balloon

Guiderope

Instrument basket

Ballast

The Valve

There exist many kinds of valves among which the two principal types are the French, with a double clack and the German, with a single clack. Those constructions are authorized where the ventilator is lifted up by pulling on the valve cord and where a spring or rubber band exerts pressure on the base of the valve thus ensuring the automatic airtight closure when the cord is released. There must be some guarantee of safety when the cord is released abruptly. The interior diameter should be the same for inflation with lighting gas or with hydrogen, and must be at least 1/30th of the maximum diameter of the envelope. The length of the valve must correspond at least to one quarter of the diameter of the valve, or otherwise the basis of the outflow volume must be at least: 0.125 m²

Note: — The safety factor of a rope is a ratio where the numerator represents the breaking point of a rope, i.e. the maximum load, and where the denominator represents the effective load that is to be applied to the rope. For example, if a rope of ·112 in diameter reaches its breaking point at 198 lbs, and if for safety reasons there should be applied a safety factor of 15, we find the effective load that can be applied to it as follows:

$$\frac{\text{breaking point load}}{\text{effective load}} \quad \text{or} \quad \frac{198 \text{ lbs}}{x} \;=\; 15 \text{ (safety factor)}$$

i.e. $\dfrac{198 \text{ lbs}}{15} = X$ or 13.2 lbs.

When calculating the resistance of ropes in new nets, the resistance of unknotted ropes must be multiplied by 20.

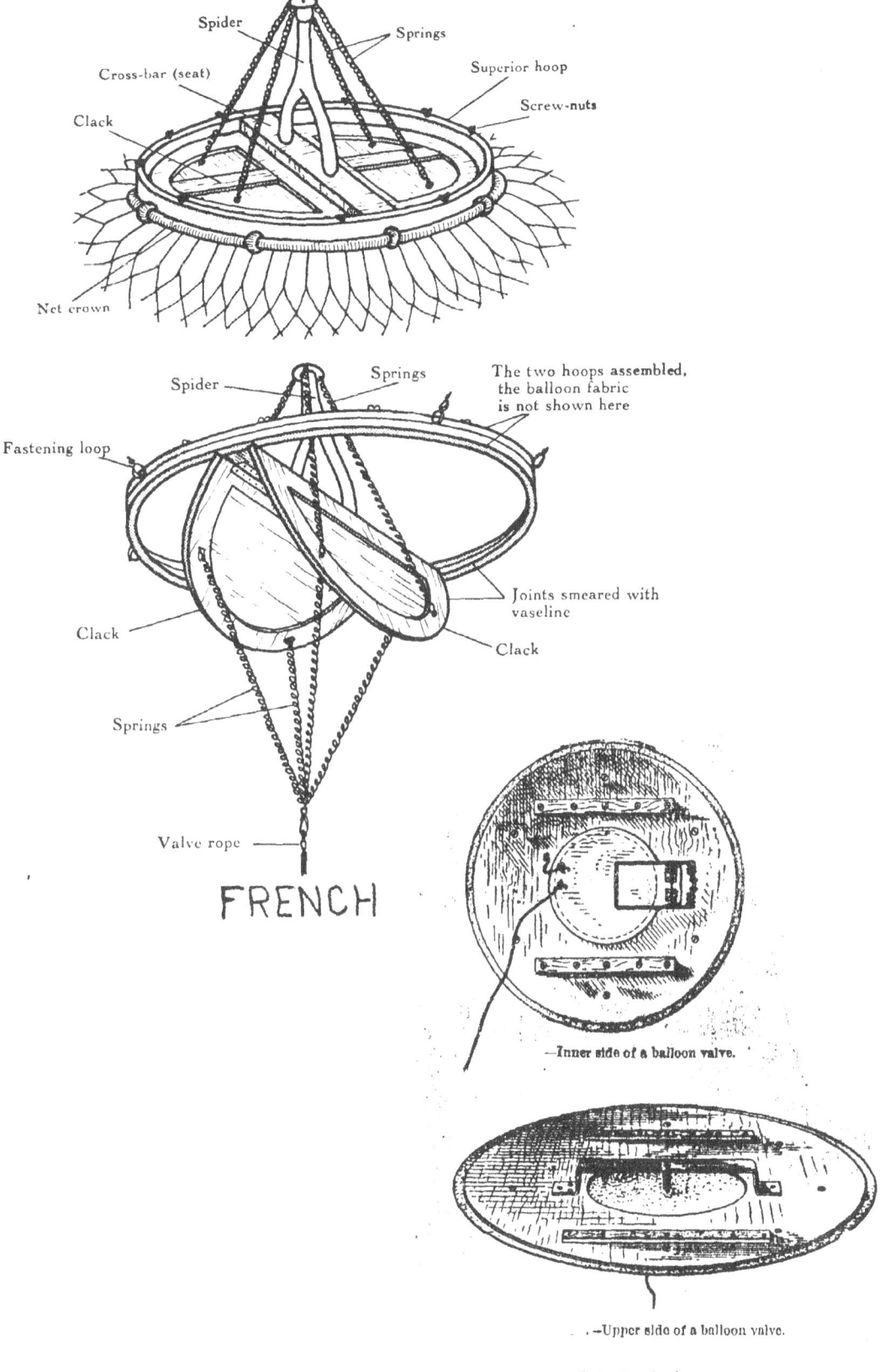

Spider

Springs

Cross-bar (seat)

Superior hoop

Screw-nuts

Clack

Net crown

Spider

Springs

The two hoops assembled, the balloon fabric is not shown here

Fastening loop

Joints smeared with vaseline

Clack

Clack

Springs

Valve rope

FRENCH

—Inner side of a balloon valve.

.—Upper side of a balloon valve.

GERMAN

CHAPTER XVI.

BALLOON CONSTRUCTION AND THE PREPARATION OF THE GAS.

BALLOONS can be filled either with hydrogen, water gas, or coal gas. The preparation of hydrogen can be effected in various ways. The method originally suggested by Charles is probably the simplest, and consists in the addition of dilute sulphuric acid to iron. But practically it leads to difficulties. The newly-generated gas is very hot, and adulterated with a certain amount of acid vapours. It must therefore be cooled and washed free from impurities. This is done by allowing it to pass through flowing water, after which it is dried by coming into contact with substances which easily absorb moisture, such as calcium chloride. It is then ready to be passed into the balloon. This method is still employed with various modifications; iron can, of of course, be replaced by zinc, and sulphuric by hydrochloric acid.

The chemical formula showing the reaction is as follows, viz.:

$$H_2SO_4 + Fe = H_2 + FeSO_4,$$

i.e., the addition of sulphuric acid to iron forms hydrogen and ferrous sulphate. From this formula it is possible to calculate the amount of gas that is formed. The atomic weights are $H = 1$, $S = 32$, $O = 16$, $Fe = 56$. A cubic foot of hydrogen weighs 0·09 oz. Suppose it is required to know how much iron and sulphuric acid will be needed to generate sufficient hydrogen to fill a balloon of 20,000 cubic feet capacity. We find, first of all, the weight of the hydrogen, which is $20,000 \times 0·09$ oz., i.e., 1 cwt. The amount of iron will be 28 times the weight of the hydrogen, and will therefore amount to 1 ton 8 cwt.; the weight of sulphuric acid will be 49 times that of the hydrogen, and is consequently 2 tons 9 cwt. In the process of the work losses of one kind or another are sure to arise, added to which the iron will probably be rusty, and the sulphuric acid will certainly contain impurities. It will therefore be found that in actual working about 20 per cent. more sulphuric acid and iron will be required than is allowed for in the calculations.

If hydrogen is generated on this system it starts very fast, but gradually the evolution of the gas becomes slower, until it finally ceases altogether, owing to the formation of a film of ferrous sulphate on the surface of the iron. The so-called circulation system was therefore introduced as an improvement, by which the fluids are kept in a state of circulation, and the iron sulphate is steadily removed in consequence. It is very important to use pure sulphuric acid, because the cheaper kinds contain arsenic. The use of impure acid has led to several fatal accidents, and the smallest amount of arsenic

0.075 pound, a cubic foot of hydrogen will lift 0.07
pound, and 1,000 cubic feet will lift 70 pounds; 1,000
cubic feet of coal gas, about 35 pounds.

It is necessary that balloons should be made to weigh
as little as possible. On this account the form becomes an object of some consideration. A spherical
form has been mathematically demonstrated to be the

Diameters.	Surfaces.	Capacities.	Ascensive Powers.	
			Hydrogen.	Coal Gas.
24	1810	7288	452	253
25	1963	8181	511	286
26	2124	9208	575	322
27	2290	10306	644	360
28	2463	11494	718	402
29	2642	12770	798	447
30	2827	14137	884	495
31	3019	15598	975	540
32	3217	17157	1072	600
33	3421	18817	1176	658
34	3630	20580	1286	720
35	3848	22449	1403	769
36	4070	24430	1527	854
37	4301	26522	1658	928
38	4536	28731	1796	1000
39	4778	31060	1942	1087
40	5026	33510	2094	1175
45	6369	47713	2982	1500
48	7200	58000	3575	2030
50	7854	65450	4091	2290
55	9503	87114	5445	3049
60	11310	113098	7069	3955
65	13273	143794	8987	5043
70	15394	179595	11225	6286

FEDERAL AVIATION AGENCY

1A14

Revision 2
GOODYEAR
S-30
S-121
813

May 1, 1965

AIRCRAFT SPECIFICATION NO. 1A14

Holder of Type Certificate: Balloon Club of America
                               230 Rutgers Avenue
                               Swarthmore, Pa.

I - Free Balloon Models S-30 and S-121, Approved May 31, 1956.

Description		
	(a) Diameter	53.4 ft.
	(b) Displacement	80,000 cu. ft.
	(c) Lifting gases	Helium, hydrogen, illuminating gas and coke oven gas.
	(d) Ballast provisions	Sand bags
	(e) Empty Weight	1205 lbs. (excluding occupants and ballast)

Number of Occupants   6 Crew and passengers.
Serial Nos. eligible   P-1 and P-3 only.

II - Free Balloon Model 813, Approved May 31, 1956.

Description		
	(a) Diameter	40.6 ft.
	(b) Displacement	35,000 cu. ft.
	(c) Lifting gases	Helium, hydrogen, illuminating gas and coke oven gas.
	(d) Ballast provisions	Sand bags
	(e) Empty weight	750 lbs. (excluding occupants and ballast)

Number of occupants   6 Crew and passengers.
Serial Nos. eligible   P-2, P-5, P-7 and P-8 only.

III - Free Balloon Model S-94 (Approved October 23, 1957).

Description		
	(a) Diameter	33.0 ft.
	(b) Displacement	19,000 cu. ft.
	(c) Lifting gases	Helium, hydrogen, illuminating gas and coke oven gas.
	(d) Ballast provisions	Sand bags
	(e) Empty weight	525 lbs. (excluding occupants and ballast)

Number of occupants   3 (crew and passengers).
Serial Nos. eligible   P-4 and P-6 only.

Specifications pertinent to all Models.

Certification basis   Type Certificate No. 1A14 (CAR 3.10)

Operating limitations (a) All flights shall be in accordance with VFR (day) rules.
                       (b) Adequate and acceptable precautions will be taken during inflation and
                           deflation procedures for the proper protection of personnel and
                           property.

. . . END . . .

Volumes of Spheres

D	V	D	V	D	V	D	V
1	0.5236	26	9203	51	69456	76	229847
2	4.189	27	10306	52	73622	77	239040
3	14.137	28	11494	53	77952	78	248475
4	33.51	29	12770	54	82448	79	258155
5	65.45	30	14137	55	87114	80	268083
6	113.1	31	15599	56	91952	81	278262
7	179.59	32	17157	57	96967	82	288696
8	268.08	33	18817	58	102160	83	299387
9	381.70	34	20580	59	107536	84	310339
10	523.599	35	22449	60	113097	85	321555
11	696.91	36	24429	61	118847	86	333038
12	904.78	37	26522	62	124788	87	344791
13	1150.35	38	28731	63	130924	88	356818
14	1436.76	39	31059	64	137258	89	369121
15	1767.15	40	33510	65	143793	90	381704
16	2144.66	41	36087	66	150533	91	394569
17	2572.44	42	38792	67	157479	92	407720
18	3053.63	43	41630	68	164636	93	421160
19	3591.36	44	44602	69	172007	94	434893
20	4188.79	45	47713	70	179594	95	448921
21	4849	46	50965	71	187402	96	463247
22	5575	47	54362	72	195432	97	477874
23	6371	48	57906	73	203689	98	492807
24	7238	49	61601	74	212175	99	508047
25	8181	50	65450	75	220893	100	523599

Ballonfabrik · D-8900 Augsburg 31 · Postfach 280

Mr. Robert J. Rechs
DELTA WIND KITES
P.O.Box 483

Van Nuys, Ca. 91408
U.S.A.

Sie schrieben am	mit dem Zeichen	Unsere Zeichen	Tag
		GL/HH/mk	22.5.1974

Dear Mr. Rechs,

We thnak you for your information and are pleased to offer as
follows:

                780 cu.m. balloon for 4 persons   DM 19.800,-
              1.260 cu.m. balloon for 6 persons   DM 29.000,--

These prices include:

Envelope with all lines, sewn-in rip panel, covered by adhesive
tape, auxiliary rip panel, appendix ring, Poeschel-ring, appendix
closing device, net, valve, bag for valve, load ring, packing
canvas for envelope, drag rope 50 m long, 30 mm dia., 1 handling
line 20 m long, 15 mm dia.

Basket:	4 persons	6 persons
Dimensions	125 x 105 cm	135 x 115 cm

Price includes 4 pockets, leather reinforcement of 4 bottom
corners, basket cover, sand trough with shovel, apron for
drag rope

                DM 2.200,-   DM 2.300,-

Additionally:
Leather upholstery of top rim of basket:
                DM   460,-   DM   510,-

Accessories:

60 - 100 sand bags à DM 13,80
Filling tube (connection between gas source and appendix) DM 33,-/m
              (standard length 25 m)
Filling socket (connection between filling tube and
              appendix)   DM 46,-

Certificate for airworthiness for export   DM 250,-

Es gelten ausschließlich unsere Zahlungs- und Lieferungsbedingungen, die auf Wunsch zugesandt werden

Telefon   (08 21) *29 395   Banken Stadtsparkasse Augsburg 084 103 (BLZ 720 500 00)
Telex   05 3 626   Landeszentralbank Augsburg 68/817 (BLZ 720 081 70)
Postscheck München 451 72   Deutsche Bank 08/700 59 (BLZ 720 700 01)

Schlauchboote - Rettungsinseln - Schwimmwesten
Faltbehälter - Freiballone - Werbeballone - Gasspeicherballone
Schutzbekleidung für See- und Luftfahrt
Traglufthallen - Flexible Konstruktionen

- 2 -

Our prices include delivery ex works, accessories packed in the basket, balloon envelope packed in envelope cover. If seaworthy packing is required, that would be charged at the nominal costs.

Our prices do not include the inscription of the balloon, since this depends very much on the design required. If you could let us have a drawing, we would be pleased to quote for this too.

Our standard balloon colours are white, yellow, orange, red and blue. A combination of these colours can be arranged.

<u>Terms of Payment:</u>	1/3 togehter with your order, rest cash against documents
<u>Time of delivery:</u>	Abt. 3 months

We would be pleased if you would join the beautiful sport of ballooning in a BFA free balloon. Looking forward to your news, we remain,

Yours faithfully,
BALLONFABRIK
See- und Luftausrüstung
GmbH + Co KG

GAS BALLOON TERMINOLOGY

APEX   is the very top of the balloon

APPENDIX is a snout like opening on the bottom of the envelope for venting
expanded gas.

APPENDIX TIE-OFF refers to the prevention of gas escaping thru the appendix
before take-off.

BALLAST is sand used to control climbs, descents and landings.

DRAG ROPE is a heavy rope tapered and weighted at one end that is used as
recoverable ballast.

ENVELOPE is the rubberized fabric or plastic material that encloses the gas.

EQUILIBRIUM is that point when lift equals weight and the balloon is
neither climbing nor descending.

FALSE LIFT refers to the venturi effect of the wind that causes the
balloon to lift before true equilibrium is reached.

SKIN TEMPERATURE refers to the temperature of the rubberized fabric envelope.

STEP CLIMB is a series of climbs and level-offs.

SUPER HEAT refers to the temperature of the gas exceeding the temperature
of the ambient air.

TRACK is the path an aircraft travels over the ground calculated in degrees
relative to true north.

WEIGH-OFF refers to the technique of determining buoyancy.

(The following is a list prepared by the National Lighter-Than-Air instructors to aid in balloon flying techniques.)

THUMB RULES

1) Lift of a balloon varies as the volume if all other conditions affecting lift remain constant.

2) Lift of a given volume of gas will decrease if atmospheric temperature increases and will increase if temperature is decreased. (Outside air is less dense at higher temperature.)

3) Where air and gas temperature change an equal amount there is no change in equilibrium if the gas is free to expand.

4) A balloon rising from the ground in equilibrium will be in equilibrium at any altitude below pressure height (if no weight is lost or gained and the superheat value does not change).

5) A balloon in equilibrium at any altitude will be in equilibrium on the ground (providing no weight is lost or gained and the superheat value is not changed in descending).

6) Atmospheric temperature will decrease approximately 1° F for every 300 feet ascent (or 3-1/3° for every 1000 feet ascent).

7) Gas volume is changed 1% for every 5° F change in gas temperature.

8) Gas density is changed 1% for every 5° F change in gas temperature.

9) Gross lift is changed 1% for every 5° F change in superheat in flight if the gas is free to expand.

10) At pressure height the gross lift will change only 1% for 45° F superheat with H_2, 1% for 20° F superheat with He, 1% for 5-10° F superheat with coke gas. This shows the danger in going from max. daytime superheat conditions to zero superheat value at night if not properly understood and counteracted.

11) 5° F superheat will lower the pressure height 360 feet below 7000 feet and 400 feet above 7000 feet. (Gas expands and fills balloon at lower altitude).

12) Lift of a given volume of gas increases if (atmospheric) barometric pressure increases and lift decreases if pressure decreases. (Density of air increases).

13) There is no change in equilibrium due to a change in barometric pressure when the gas is free to expand.

14) The higher the atmospheric humidity the less the lift. (Water takes up air space and weighs less.)

15) Barometric pressure will decrease approximately 1" for every 1000 ft. of ascent in the lower atmosphere. (Std. = 30" Hg).

16) At pressure height 1% change in gas density or sp. gr. for H_2 change gross lift 0.1%, for He changes gross lift 0.2%, and for coke gas changes gross lift approximately 0.3%.

17) In ascending under average atmospheric conditions, the volume will increase 1% for every 360 feet ascent below 7000 feet and 1% for every 400 feet above 7000 feet.

18) In ascending above pressure height, gross lift is reduced 1% for every 360 feet below 7000 feet and 1% for every 400 feet above 7000 feet.

19) 1% of mass of gas is lost going 360 feet over pressure height below 7000 feet and 1% for 400 feet above 700 feet.

20) At pressure height, if 1% of gross lift is thrown over as ballast, equilibrium will be reached when 1% of the gas has been valved (350 feet^3).

GASSING OFF (Cont. from Page 2)

Metal license plates bearing the name Wingfoot Lighter-Than-Air Society and depicting LTA operations may be obtained by mailing one dollar ($1.00) to: Wingfoot LTA Society, Dept. 450-G, 1210 Massillon Road, Akron 15, Ohio. If plates are to be mailed, enclose additional 25¢ for postage and handling.

Circle: A = Area
        D = Diameter
        r = Radius
        π = 3.1416

$A = \pi r^2$
$C = \pi D$

Cone: S = slant height
       V = Volume
       C = Circumference
       B = Base Area
       h = Height

  frustrum

    P_1 = perim. circum; top

    P_2 = perimeter circum: Bot.

    Vf = Volume of Frustrum
    Af = Area
    B = Base area

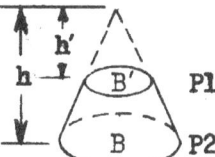

$S = \sqrt{r^2 + h^2}$
$A = \frac{1}{2} SC$
$V = 1/3\ Bh$
$B = \pi r^2$

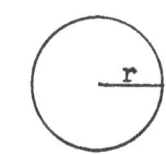

$Af = \frac{1}{2} S\ (P_1 + P_2)$
$Vf = 1/3\ Bh - 1/3\ B'h'$

Sphere: r = radius
        A = area
        V = volume

  frustrum
    h = height
    D = diameter
    A - area

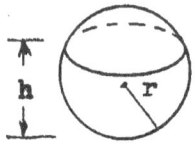

$D = \sqrt[3]{6V}$
$A - 4\ r^2$
$V - 4.188 r^3$

$Vf = 4/3\ \pi r^3$
$Af - 2\pi hr$
$Vf = 4.1888\ r^3$

Note: π = 3.14159 $4/3\ \pi$ = 4.1888
     $2\ \pi$ = 6.2832 $4\ \pi$ = 12.5664
     $3\ \pi$ = 9.4248 $5\ \pi$ 15.708

Distance - Linear   Volume - Capacity

  1 meter = 39.37 in.   cu meter = 1.308 cu. yds.
    "   = 3.281 ft.   "   = 35.314 cu. ft.
  1 foot = 30.48 centimeters   cu. ft. = 1728 cu. in.
    "   = .3048 meters   "   = .028 cu. meters

Surface - Area

  1 sq. meter = 1.196 sq. yds.   sq. yd. = .836 sq. meter
             10.764 sq. ft.   sq. ft. = .092 sq. meter

To Get:
    Cubic meters, multiply cubic ft. times   .028 $1. °C = F - 32$

    cubic feet, multiply cubic meters times 35.314

    feet, multiply meters times 3.281

    meters, multiply feet times .3048

    yards, multiply meters times .9144

    pounds, multiply kilograms times 2.2

    kilograms, multiply pounds times .4535

APPENDIX
CONVERSION UNITS
Applicable to balloons

```
LENGTH       (conversions)                              (equivilants)
  INCHES     x   2.54    = Centimeters !  1 Inch   =  2.54     c.m.
    "        /  39.36835 = Meters      !  1 Inch   =  .0254    Meters
  FEET       x    .3048  = Meters      !  1 Foot   =  .3048    Meters
    "        /   3.28075 = Meters      !  1 Foot   =  30.48    Centimeters
  YARDS      x    .9144  = Meters      !  1 Yard   =  .9144    Meters
    "        /   1.0936  = Meters      !  1 Meter  = 3.2808    Feet
  METERS     x   3.2808  = Feet        !  1 Meter  = 39.36835  Inches
    "        /    .3048  = Feet        !  1 Meter  = 1.0936    Yards
  METERS     x   1.0936  = Yards       !  1 Sq.Met.=  10.76    Sq.Ft.
    "        /    .9144  = Yards       !  1 Sq.Met.=  1.196    Sq.Yd.

DISTANCE:   (1 Nm = 6080.2 Feet)      (1 Kilometer =3280.8 Feet)
  S-MILES    x    .62137 = K-Miles    !  1 S-Mile = 1.6093    K-Miles
  S-MILES    x    .53996 = N-Miles    !  1 S-Mile =  .86898   N-Miles
  N-MILES    x   1.15078 = S-Miles    !  1 N-Mile = 1.15078   S-Miles
  N-MILES    x   1.8520  = K-Miles    !  1 N-Mile = 1.852     K-Miles
  KILOMETER  x    .62137 = S-Miles    !  1 Kilometer=.53996   N-Miles
  KILOMETER  x ? .53996  = N-Miles    !  1 Kilometer = .6214  S-Miles

WEIGHT:                                (1 Pound    = 453.6  Grams)
  POUNDS     x 2.2046229= Kilograms !  1 Kilogram = 2.2046 Pounds
  KILOGRAMS  / .45359237= Pounds    !  1 Pound    = .45359 Kilogram

VOLUME:     (1 Cu.Foot = 1728 C.I.)  (1 Cu.Yard = 46656. Cu.Inches)
  GALLONS    x    .133681= Cu.Feet  !  1 Gallon = 231.0163 Cu.Inches
    "        /   7.48052 = Cu.Feet  !  1 Gallon =   .13369 Cu.Feet
  LITERS     x    .26418 = Gallons  !  1 Gallon = 3.7854   Liters
    "        /   3.7854  = Gallons  !  1 Liter  =  .26418  Gallons
  Cu.METERS  x    .031897= Cu.Feet  !  1 Cu.Foot = 7.48055 Gallons
    "        /   7.48055 = Cu.Feet  !  1 Cu.Foot = .031897 Cu.Meters
  Cu.FEET    x 35.31     = Cu.Meters !  1 Cu.Yard =  .7646  Cu.Meters
    "        /    .02832 = Cu.Meters !  1 Cu.Meter = 35.31  Cu.Feet
  Cu.YARDS   x   1.308   = Cu.Meters !  1 Cu.Meter =1.308   Cu.Yards
  1 Barrel petrol = 42 Gal.          !  1 Cu.Meter =264.17  Gallons

TEMPERATURE: (1 Degree F =1.8C + 32)  1 Degree C = (F-32)/1.8
   CENTIGRADE = 5/9 (F + 32)       !  FAHRENHEIT = 9/5C + 32

LIQUID PETROLEUM GAS:                  (LPG = 6 Lb/Gal.@ 59F)
  PROPANE = 1925 BTU/Lb. and has a Molecular Wt. of 44.09
   50/50  = 1975 BTU/Lb.  "  "  "  "  "     "   "  51.10
  BUTANE  = 2025 BTU/Lb. and has a Molecular Wt. of 58.12
  Propane (C3H8) boils at -44 Degrees F° (
          reaches 60 PSI at 35 F°, 100 PSI at 65 F°.
  Butane (C4H10) boils at 32 Degrees F.
          reaches 60 PSI at 120F, 120 PSI at 212F.
  1 Gal. LPG @ 60F°= 1.07 Gal. @ 100F°= .92 Gal. @ 0F°.
  10 Gal. BUTANE @ 60F° looses 26% BTU at 212F°
          if relief valve pressure stays constant at 250 PSI.

HELIUM:  76 lbs. lift/1000 C.F.     |(Hydrogen = 79 lbs/1000)
  1 C.F. liquid = 7.6 Pounds        | 1000LL = 268.39 Pounds
  1 C.F. liquid = 682 C.F. gas      | 1000LL = 768.231 C.M. Gas
  1 C.F. liquid = 47 lbs. lift      | 1000LL = 1660#(753K) Lift
```

www.ingramcontent.com/pod-product-compliance
Lightning Source LLC
Chambersburg PA
CBHW080541190526
45169CB00007B/2587